The New Iraqi Journal of Medicine: Volume three (2007)

Editor: Aamir Jalal Al Mosawi

Professor, Aamir Jalal Al Mosawi,
Founding editor
The New Iraqi Journal of Medicine
   1-Advisor Doctor
   Training and Development Centre
   Iraqi Ministry of Health
   2-Head Iraq Headquarter of Copernicus Scientists
International Panel
   Baghdad, Iraq.
   E-mail:almosawiAJ@yahoo.com

The New Iraqi Journal of Medicine: Volume three (2007)

Editor: Aamir Jalal Al Mosawi

ISBN: 978-1-312-56508-1

INDEX COPERNICUS
SCIENTISTS
Iraq Headquarter
Copernicus Scientists International Panel

Content

Content (Cont.)

Content (Cont.)	
Ghazi Aboud, Mohammad Rasheed	
Letter to the editor: Sexual dysfunction in uncomplicated diabetic men: Positive association with insulin therapy. The New Iraqi Journal of Medicine 2007; 3 (3): 77-78.	**110-114**

Preface

The New Iraqi Journal of Medicine

The benefit of scientific medical peer-reviewed writing and publication is to document scientific facts, practices, and new hypothesis and to apply proven facts to patient care by challenging the current practices and subjecting them to the judgment of peers. In this way the scientific knowledge is advanced contributing to improved patient care, medical and health practices.

There has been a tremendous need for an independent peer-reviewed medical journal that aims at the advancement of medical knowledge rather than promotion in Iraq.

The New Iraqi Journal of Medicine was founded in 2005 and within the same year it became the official journal of the Iraqi Ministry of Health. Soon, the journal was accredited by Iraq headquarter of Copernicus Scientists International Panel. Later, the journal became the official journal of the Iraqi Ministry of Health and Iraq headquarter of Copernicus Scientists International Panel. The journal was the first Iraqi medical journal to be listed and indexed by international index which is the Copernicus Journal Master list.

From the first beginning of this journal in 2005, the decision was made that the journal should adhere to the internationally accepted standards of modern peer-reviewed medical journalism. All the editorial policies were adapted from the policies of the International Committee of Medical Journal Editors (ICMJE), World Association of Medical editors (WAME), and the Committee on publication ethics (COPE). Initially, the primary goal of the Journal was to publish research works confined to Iraq, which were of importance for both local and international readers. Another aim was to make the research works easily available for international readers. Thus a sincere attempt was being made to disseminate the medical knowledge to all medical fraternity of the world.

The journal had editors from 4 major continents in the world. In addition it had an International editorial board of international editors including some editorial board members who have the privileges of the Editor-in-Chief, as they can receive, evaluate manuscripts and provide the journal with an editorial decision.

The journal was successful in publishing papers in more than 20 areas and disciplines of medicine. Authors located in Iraq, USA, UK, Italy, Jordan, Germany, India, Malaysia, Iran, Qatar, Malawi, and Zambia contributed to the journal.

Nine volumes of this international journal were published, but it was finally stopped because of the lack of resources and qualified staff.

The aim of this book is to publish the important scientific articles and research published in issues of "The New Iraqi Journal of Medicine" Volume three, 2007. In this volume (three-2007), authors from Iraq, India, USA, Jordan, Malawi, Zambia, and Malaysia contributed to the journal in many fields including anatomy, psychiatry, cardiology, public health, hematology, and general surgery.

The prevalence of violence among a group of married women attending two teaching hospitals in Baghdad

The New Iraqi Journal of Medicine 2007; 3 (1): 8-13.

Maha Adnan Abdul Jabbar
University Hospital in Al Kadhimiyia
Baghdad Iraq

Abstract

Background: Violence is an important public health problem world wide. Domestic violence against women is particularly concerning as its generally perpetrated within relationships that are supposed to involve care and protection is a much more serious problem than violence perpetrated by strangers. **The aim of this study** is to assess the prevalence of violence against women, and studying possible risk factors that may be associated with this.

Patients and methods: From March 2006 to August 2006, 323 Arabian women 265 married (82%), 34 widows (10.5%), 24 divorced (7.4%) were observed at Al-Kadhimiyia and Al-Kindy teaching hospitals and interviewed to determine their exposure to violence. Their age ranged from 18 to 70 years (Mean age 39 years).They were interviewed regardless whether they were attending as a patients or not.

Results: Domestic violence (DV) by husband was reported in 186 women (57.6 %) for at least one incident and 117 women (44%) for continuous or current violence. Violence not by husband was reported in 67 women (20.7%) as past experience. While 16 women (5%) with continued exposure to violence. Psychological violence was the major type of DV by husband occurring in 185 women (57.3%), followed by physical occurred in 128 women (39.6%) and finally sexual violence (rape) which occurred in 48 women (14.9%). Simple assault (slapping or hitting) was the major form of physical violence by husband occurred in 85 cases (66.4%).

Conclusions: About half of women were victims of DV in this study. Psychological violence was the major form. Some women were at higher risks for DV by husband. Reliable and accurate data regarding violence against women are still limited and information about different incidents of violence against women is still confidential and not reported to governmental institutions or agencies especially in this country. For this reason, further sophisticated studies about this subject are recommended.

Introduction

Violence is an important public health problem world wide. Despite the prevalence of male victimization, women are overrepresented in virtually all forms of research on victimization. Women violence perpetrated within relationships that are supposed to involve care and protection is a much more serious problem than violence perpetrated by strangers [1].

In one study about (29%) of women and about (22%) of men had experienced sort of physical, sexual, or psychological intimate partner violence during their life time[2].**The aim of this study** is to assess the prevalence and type of violence against women, and studying possible risk factors that may be associated with this.

Patients and methods: From March 2006 to August 2006, a convenient sample(WHO sampling method)[3] of 323 Arabian women (265 married, 34 widows, 24 divorced) were observed at Al-Kadhimiyia and Al-Kindy teaching hospitals and interviewed to determine their exposure to violence. Their age ranged from 18 to 70 years (Mean age 39 years).They were interviewed regardless whether they were attending as a patients or not. The data was gathered using a questionnaire form constructed by the researcher, including past or present history of experiencing violence by husband and or family, type of violence, with some female criteria. Mentally retarded were excluded.

Results: Domestic violence (DV) by husband was reported in 186 women (57.6%) for at least for one incident, continuous or current violence occurred in 117 women (44%).Violence not by husband was reported in 67 women (20.7%) including 16 women (5%) with continued exposure to violence. Psychological violence was the major type of DV by husband occurring in185 women (57.3%), followed by physical occurred in 128 women (39.6%) and finally sexual violence (rape) which occurred in 48 women (14.9%).

Psychological violence by husband significantly led to physical violence ($P<0.05$) in 127 cases of violence (68.6%) and the last significantly led to sexual violence ($P<0.05$) in 44 cases of violence (34.4%). About (2.5%) of women of study group had family history of first and second degree relative who was murdered by her husband, and (2.8%) had family history of a relative (women) who were killed or kidnapped not by husband. Police report took place only in 12 cases (9.1%) for DV by husband while no police report recorded for violence not by husband.

Table (1) shows the distribution of violence, types of violence, murdered females, and police report against offender(s).

Simple assault (slapping or hitting) was the major form of physical violence by husband occurred in 85 cases (66.4%).Aggravated assault (using knife or gun) occurred in 5 cases (3.9%). DV by husband was associated with bruises in 61 cases (47.7%), permanent deformity in 14 cases (10.9%), and psychological disturbances like anxiety, low self esteem, depression, and suicide in 132 cases (71.9%). Other forms of violence include son preference occurred in 57 cases (18.1%) and work harassment occurred in 9 cases (14.8%). Table (2) shows the distribution of sub-types of violence.

Psychological violence	Violence by husband	Violence not by husband
Call by names	21(11.3 %)	7(10.6%)
Humiliation	33(17.7%)	21(31.8%)
All types	132(71%)	38(57.6%)
Total	186	66
Physical violence		
Slapping/ hitting	85(66.4%)	27(93.1%)
Kicking by foot	2(1.6%)	0
Using object	2(1.6%)	0
All types	34(26.6%)	2(6.9%)
Using knife/ gun	5(3.9%)	0
Total	128	29
Sexual violence		
Rape	48(14.9%)	0
Sexual assault	0	3(0.9%)

Table (1): Distribution of violence, types of violence, murdered females, and police report against offender(s).

Psychological violence	Violence by husband	Violence not by husband
Call by names	21(11.3 %)	7(10.6%)
Humiliation	33(17.7%)	21(31.8%)
All types	132(71%)	38(57.6%)
Total	186	66
Physical violence		
Slapping/ hitting	85(66.4%)	27(93.1%)
Kicking by foot	2(1.6%)	0
Using object	2(1.6%)	0
All types	34(26.6%)	2(6.9%)
Using knife/ gun	5(3.9%)	0
Total	128	29
Sexual violence		
Rape	48(14.9%)	0
Sexual assault	0	3(0.9%)

Table (2) shows the distribution of sub-types of violence.

By studying female's criteria of those women experienced DV by husband. Violence by husband occurred in all 24 divorced women during the period of their marriage making (100%). Violence also occurred in (90%) of the 41 women got married with out their agreement, and in (90.5%) of the 21 women whose husbands are married to more than one wife. For years of formal education for woman highest violence rate was (75.3%) for (0 - 4) years, and in the presence of son preference the rate was (73.7%) or 42 case from 57 women with history of son preference, and all

13

above were significantly associated with DV by husband (P<0.05).

Other factors that may be associated with DV by husband include the following where violence rate was, (59%) in not working women, (62.7%) in history of former violence by family and or relatives and (69.2%) for age of women at marriage at (10 -14) years. For male(husband) significant risk factors for perpetrator include, alcohol intake (79.5%), illegal relationships (95%), and years of formal education for husband (71.8%) for (0 - 4) years (for all P<0.05). Other factors that may be associated, but not significantly, with perpetration in male include, not working man (88.8%), drug addiction (100%), and later age at marriage (65.2%) for (>38 years) at marriage. Also there are some factors that may be associated with DV against women include residence (61.9%) for rural, income (62%) for not enough income, home ownership (65.6%) for nor own nor rent house. Table (3) summarizes some female factors associated with violence.

Table (4) shows the relation between years of formal education for women and husbands / Age at marriage for women &husbands and violence. Table (5) summarizes some male and other factors associated with violence against women.

It is interesting to know that (33.4%) of women accept DV by men, and (23.5%) believe that woman should not argue man in discussion, beside (38.4%) think that man is better than woman.

Table(3)	Violence within marriage				Total		X²	P-value
Current marital status	yes	%	no	%	No.	%		
Married	145	54.7	120	45.3	265	100	19.371	0.000
Widow	17	50	17	50	34	100		
Divorced	24	100	0	0	24	100		
Total	186	57.6	137	42.4	323	100		
Occupation of women								
Employee or free job	45	53.6	39	46.4	84	100	0.749	0.387
Housewife	141	59	98	41	239	100		
Total	186	57.6	137	42.4	323	100		
Married by agreement								
By agreement	149	52.8	133	47.2	282	100	20.507	0.000
No agreement	37	90	4	9.8	41	100		
Total	186	57.6	137	42.4	323	100		
Multiple wives for husband								
Present other wife	19	90.5	2	9.5	21	100	9.948	0.002
No other wife	167	55.3	135	44.7	302	100		
Total	186	57.6	137	42.4	323	100		
Son preference								
Son preference	42	73.7	15	26.3	57	100	7.494	0.006
No preference	139	53.9	119	46.1	258	100		
Total	181	57.5	134	42.5	315	100		
Former violence not by husband								
Former violence	42	62.7	25	37.3	67	100	0.901	0.343
No former violence	144	56.3	112	43.8	256	100		
Total	186	57.6	137	42.4	323	100		

Table 3: Female factors associated with violence

Table 4	Violence within marriage				Total		X²	P-value
Years of formal education for women	yes	%	no	%	No.	%		
0 - 4	67	75.3	22	24.7	89	100	16.365	0.001
5 - 9	65	53.3	57	46.7	122	100		
10 - 14	38	48.1	41	51.9	79	100		
> 15	16	48.5	17	51.5	33	100		
Total	186	57.6	137	42.4	323	100		
Age of women at marriage								
10 - 14	27	69.2	12	30.8	39	100	6.615	0.158
15 - 19	72	61	46	39	118	100		
20 - 24	45	50	45	50	90	100		
25 - 29	27	61.4	17	38.6	44	100		
> 30	15	46.9	17	53.1	32	100		
Total	186	57.6	137	42.4	323	100		
Years of formal education for husbands								
0 - 4	28	71.8	11	28.2	39	100	8.015	0.046
5 - 9	86	61.9	53	38.1	139	100		
10 - 14	46	50	46	50	92	100		
> 15	26	49.1	27	50.9	53	100		
Total	186	57.6	137	42.4	323	100		
Age of husbands at marriage								
14 - 19	22	59.5	15	40.5	37	100	2.066	0.724
20 - 25	49	52.7	44	47.3	93	100		
26 - 31	76	60.3	50	39.7	126	100		
32 - 37	24	54.5	20	45.5	44	100		
>38	15	65.2	8	34.8	23	100		
Total	186	57.6	137	42.4	323	100		

Table 4.

	Violence within marriage				Total		X^2	P-value
Occupation of husband	yes	%	no	%	No.	%		
Employee	67	58.3	48	41.7	115	100		
Free job	105	55.3	85	44.7	190	100	3.446	0.179
Not working	14	88.8	4	22.2	18	100		
Total	186	57.6	137	42.4	323	100		
Alcohol intake for husband								
Drink alcohol	31	79.5	8	20.5	39	100		
Not alcoholic	155	54.6	129	45.4	284	100	8.711	0.003
Total	186	57.6	137	42.4	323	100		
Drug addiction for husband								
Drug addict	3	100	0	0	3	100		
Not addicted	183	57.2	137	42.8	320	100	2.230	0.135
Total	186	57.6	137	42.4	323	100		
Illegal relationship for husband								
Illegal relationship	19	95	1	5	20	100		
No relation	167	55.1	136	44.9	303	100	12.219	0.000
Total	186	57.6	137	42.4	323	100		
Husband relative to his wife or not								
Foreigner	87	58	63	42	150	100		
2nd degree relative	66	56.4	51	43.6	117	100	0.118	0.943
3rd degree relative	33	58.9	23	41.1	56	100		
Total	186	57.6	137	42.4	323	100		
Residence								
Urban	173	57.3	129	42.7	302	100		
Rural	13	61.9	8	38.1	21	100	0.172	0.679
Total	186	57.6	137	42.4	323	100		
Income								
Not enough	57	62	35	38	92	100		
Enough for living/ saving	129	55.8	102	44.2	231	100	1.006	0.316
Total	186	57.6	137	42.4	323	100		
Home ownership								
Own	104	57.5	77	42.5	181	100		
Rent	42	51.9	39	48.1	81	100	2.685	0.261
Neither own nor rent	40	65.6	21	34.4	61	100		
Total	186	57.6	137	42.4	323	100		

Table (5): Male and other factors associated with violence women.

Discussion

These results showed that high proportions of women are victims of any sort of violence whether psychological, physical, and or sexual from their husbands and experiencing psychological violence alone is not a simple problem because it may leads to amore sever form of violence. Lower rate of report make it difficult for researchers to investigate the problem or to collect data from governmental agencies and or police stations.

The effects of violence are very important thing as they may interfere with women's life and performance. Women's ideas and believes regarding marital relationship and women's rights should be improved in the way that every one in this relationship has rights and obligations. Certain groups of women are at higher rate of risk of being abused by their husbands include those who are not working, low educated women, married without agreement, presence of other wife, former history of violence and son preference by parents of women.

Results of this study are to some what similar to results of researches done in Arabic areas [4] where 52% of Palestinians women showed that they had been experienced physical violence at least one time for the year 2000 and in study done in Jordan showed that percent of women who were continuously being physically abused by their husbands was 47.6%. But there is a little difference from studies done nation wide. The reason may be the socio-demographic features in common for Arabic countries. Some forms of violence were prevalent in this study while other forms were absent like female circumcision.

18

Some factors might have contributed in prevailing violence in this country including successive wars, deterioration of security situation, unemployment, and psychological traumas contributed to frustration which may be expressed as verbal or physical violence. Absence of reliable data regarding these incidents of violence has faced us with difficulties in doing a research regard this subject, therefore; we depended this type of study for our research which may not reflect the true prevalence of violence in this country.

Conclusions

About half of women were victims of DV in this study. Psychological violence was the major form. Some women were at higher risks for DV by husband. Reliable and accurate data regarding violence against women are still limited and information about different incidents of violence against women is still confidential and not reported to governmental institutions or agencies especially in this country. For this reason, further sophisticated studies about this subject are recommended.

References

1-Bessel A. Vander Kolk, M.D. Physical and sexual abuse of adults. In: Benjamin J.Sadock, M.D, Virginia A. Sadock, M.D. Comprehensive text book of psychiatry. Seventh edition, 2000 ;(2), Pp: 2002-2007.

2-Coker AL, Davis KE, Arias I, Desai S, Sanderson M, Brandt HM, et al. Physical and mental health effects of intimate partner violence for men and women. American Journal of Preventive Medicine 2002; 23 (4): 260-8.

3-WHO "Sampling Method and Sample Size". In: Health research methodology: A guide for training in research. www.wpro.who.int/internet/files/pub/352/71.pdf (Accessed at June, 2006).

4-Nasser L. (2001) Violence against women and children: Finding from Amman the Arabic Center for Information and Studies.www.amanjordan.org (Accessed at June, 2006).

Anomalous veins of the face and neck: an anatomical study with clinical implications.

The New Iraqi Journal of Medicine 2007; 3 (1): 21-24.

Shipra Paul
Director Professor Department of Anatomy
Maulana Azad Medical College
Bahadur Shah Zafar Marg
New Delhi-110002.India

Srijit Das
Associate Professor, Department of Anatomy
Maulana Azad Medical College
New Delhi-110021, India

Abstract

Background: The retromandibular vein normally divides into anterior and posterior divisions at the lower part of the parotid gland. The anterior division of the retromandibular vein unites with the facial vein to form the internal jugular vein, while the posterior division joins the posterior auricular vein, to form the external jugular vein. The posterior auricular vein may sometimes exhibit variations. The present research study, reports a lower division of the retromandibular vein and the absence of posterior auricular vein, and discusses its clinical implications.

Objective: To study a case of anomalous formation of external jugular vein, in which the posterior auricular vein was absent and did not contribute to the formation of external jugular vein.

Material and methods: We observed the drainage pattern of the posterior auriculr vein on both sides of 20 human cadavers (n=40).Any anomaly pertaining to the posterior auricular vein was observed.

Results: We observed the absence of posterior auricular vein on both sides of a 58 year male cadaver. On both sides, the retromandibular vein divided into anterior and posterior divisions, 1.5 cm below the lower border of the parotid gland. The posterior auricular vein was absent.

The posterior division of the retromandibular vein continued as the external jugular vein, and no other associated venous anomalies were observed.

Conclusion

Knowledge of venous anomalies may be important for academic interest and beneficial for facial surgeons and radiologists performing angiographic studies.

Key words Posterior auricular, retromandibular, external jugular, face, vein, anomaly, variations.

Introduction

The retromandibular vein (RMV) is formed inside the parotid gland, by the union of the superficial temporal and the maxillary veins [1, 2]. The RMV divides at the lower part of the parotid gland into an anterior and posterior division [1, 2]. Both the divisions emerge from the lower part of the gland, the anterior division of which joins the facial vein (FV) at the lower border of the mandible, to form the common facial vein which drains into the internal jugular vein (IJV).

The posterior auricular vein (PAV) arises from the parieto-occipital network and descends below the auricle to join the posterior division of the RMV to form the EJV [1]. EJV is commonly used for cannulation to conduct diagnostic procedures or intravenous therapies [3]. Vascular anomalies pertaining to the EJV as such are very rare and there is also paucity of literature on the absence of PAV [4].

The topographical anatomy of the neck and the FV is important for operating surgeons and radiologists interpreting sonographic investigations. While designing the posterior auricular flap for reconstructive surgeries, pre-operative planning is essential and the anomalies pertaining to PAV may be kept in mind [5].

The increase in stent grafting procedures for treating neuro-vascular diseases also requires a thorough knowledge of the vascular anatomy of the neck region. The present study, describes an unusual case of the RMV dividing much below its usual position i.e. at a distance of 1.5 below lower border of parotid gland and the absence of PAV. There are many research reports on the facial vein and EJV anomalies, but in the present case, a lower division of RMV with absence of PAV, is a rare entity, which may be important for academic and clinical purpose.

Materials and Methods

We observed the PAV on both sides of 20 formalin fixed cadavers (n= 40). The abnormalities pertaining to the PAV were studies in detail in these twenty cadavers. The specimen was studied in detail and photographed (Figure.1).

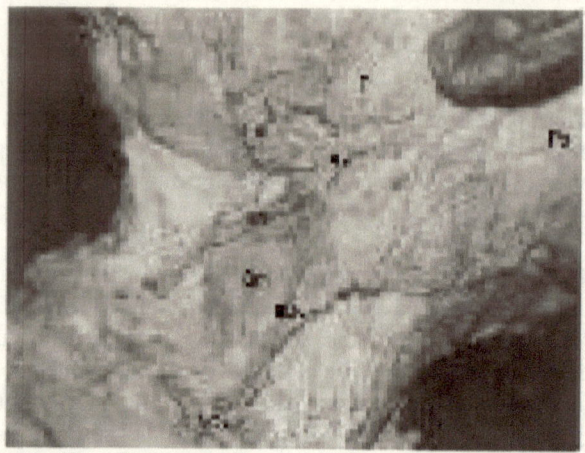

Figure (1): Demonstrates absence of the posterior auricular vein from its usual position. P (Parotid gland), **Fv** (Facial vein), **Pa** (The usual position of the posterior auricular vein, it is absent in this case), **RV** (Retromandibular vein), **Ijv** (Internal jugular vein) **Sm** (Sternocleidomastoid muscle), **EJv** (External jugular vein) **SCv** (Subclavian vein).

Results

Out of 40 cases studied, the anomalous venous pattern PAV was noted on both sides of a 58 year male cadaver (5%). The FV traversed back from the lower border of the angle of the mandible. The RMV divided into anterior and posterior branches 1.5 cm below the lower border of the parotid gland. The PAV was characteristically absent. The EJV which usually joins the posterior division of the RMV was not observed, rather the EJV was formed as a continuation of the posterior division of the RMV .No other venous anomalies were observed.

Discussion

There are research reports on the FV uniting with the RMV in the parotid gland itself [6] but the present anomaly of the RMV dividing into anterior and posterior divisions, 1.5 cm below the parotid gland, is a rare finding. A recent research study, had reported the absence of anterior division of RMV, with the facial vein continuing as the IJV [7]. In the present case, we observed the posterior division of the RMV continuing as EJV, with the absence of PAV. An unusual finding of the PAV being absent also suggests that the other veins like occipital veins may be draining a larger part.

The facial vein has also been described to drain into the EJV [8]. In the present case, the facial vein joined the anterior division of the RMV 1.5 cm below the parotid gland, thereby resulting in a much smaller IJV. The IJV is usually accessed for diagnostic and therapeutic purpose and any anomaly of this kind, has to be borne in mind.

25

The variations in these veins could be explained in terms of the regression and retention of various parts of the veins as found in the rhesus monkey [8]. Absence of PAV, as observed in the present case, may puzzle the radiologists who might be exploring the PAV for investigative procedures. As anatomists we believe, that central venous catheterization may be difficult in the presence of such anomalies. While designing flaps in the posterior auricular region, the anatomy of the PAV may be important [5]. In the absence of PAV, the flaps may not be possible because of less vascularity to the region. Surgeons have used prefabricated galeal flaps based on the superficial temporal or postauricular vessels for ear, cheek, mandible, and cranium reconstructions [9].

The use of retroauricular flaps for facial reconstructive surgeries has gained wide popularity. It has been found that the retroauricular flap provides normal color, texture, and thickness and thus is an optimal anatomic and aesthetic reconstruction with minimal donor-site morbidity [10].Hence it is very important to know the variations of the superficial venous system in order to avoid damage to the fascia and ensure the survival of any flap [11].

The frequency of cerebral developmental venous anomalies, occurring in patients with cervicofacial vascular malformation is much higher than in general population [12]. Any venous anomalies may be explored for associated malformations.

Radiologists interpreting sonographic investigations may be puzzled with venous anomalies. Often during Doppler ultrasound procedures, a correct picture of the veins helps in proper interpretation and any deviation from the usual anatomy may cause erroneous interpretation.

Conclusion

The present study highlights the absence of PAV and reports the union of the anterior division of RMV with the facial vein, much below the usual lower border of the parotid gland, which is a rare finding. Knowledge of such venous anomalies may be helpful for academic, clinical and investigative procedures.

References

1-Standing Susan: Gray's Anatomy. The Anatomical Basis of Clinical Practice. 39th edition, 2005, Philadelphia, Elsevier Churchill Livingstone; pp- 511-12 & 516.

2-Snell Richard S: Clinical Anatomy. Seventh edition, 2004, Baltimore, USA, Lippincott Williams & Wilkins; pp- 775.

3-Gupta Y, Tuli A, Chowdhury S: Facial vein terminating in the external jugular vein. An embryologic interpretation. Surg Radiol Anat 1997; 19(2):73-7.

4-Ahuja AT, Yuen HY, Wong KT, King AD, Abdullah V, To E, Chau YP, Ma KF: External jugular vein vascular malformation: sonographic and MR imaging appearances. AJNR Am J Neuroradiol. 2004; 25(2):338-42.

5-Kobayashi S, Nagase T, Ohmori K: Colour Doppler flow imaging of postauricular arteries and veins. Br J Plast Surg. 1997; 50(3):172-5.

6-Kopuz C, Yavuz S, Cumhur M, Tetik S, Ilgi S: An unusual coursing of the facial vein. Kaibogaku Zasshi. 1995; 70(1):20-2.

7-Nayak BS: Surgically important variations of the jugular veins. Clin Anat. 2006; 19(6):544-6.

8-Gupta V, Tuli A, Choudhry R, Agarwal S, Mangal A: Facial vein draining into external jugular vein in humans: its variations, phylogenetic retention and clinical relevance. Surg Radiol Anat. 2003; 25(1):36-41.

28

9-Roger KK, Joseph U: Fascia flaps in the head and neck .In Fasciocutaneous flaps. Blackwell, London, 1992, pp 27-29.

10-Ozerdem OR, Anlatici R, Sen O, Yildirim T, Bircan S, Aydin M: Prefabricated galeal flap based on superficial temporal and posterior auricular vessels. Plast Reconstr Surg. 2003; 111(7):2166-75.

11-Ninkovic M, Hubli E, Anderl H: Facial reconstruction using a retroauricular-temporal free flap. Plast Reconstr Surg. 1998; 102(4):1147-50.

12-Enjolras O, Boukobza M, Guichard JP, Gelbert F, Merland JJ: Cervicofacial superficial venous malformations and developmental abnormalities. Ann Dermatol Venereol. 1996; 123(4):235-9

Rational for the possible introduction of Haemophilus influenzae vaccine into the Iraq vaccination program

The New Iraqi Journal of Medicine 2007; 3 (1): 25-28.

Aamir Jalal Al Mosawi
Head of the department of pediatrics
University Hospital in Al Kadhimiyia

Abstract

Meningitis caused by bacteria remains one of the most potentially serious infections in infants and children with high risk of acute complications and chronic morbidity. Haemophilus influenzae was the most common bacterial pathogens causing meningitis during the first year of life before the introduction of vaccines against Haemophilus influenzae type b (Hib).The wide spread use of Hib vaccines was associated with marked decline in the frequency of Haemophilus infection, and Haemophilus influenzae is no longer the most common bacterial pathogens causing meningitis during the first year of life in many geographic areas such as UK and USA. In Iraq the precise and even the rough contribution of Haemophilus influenzae and other bacterial pathogens to acute bacterial meningitis and acute CNS infection in general remains unknown.

The aim of this paper is to discuss the rational for the possible introduction of Hib vaccine into the Iraq vaccination program based on the success associated with the introduction of the Hib vaccines into the developed counties vaccination program.

Keywords: Bacterial meningitis-Iraq vaccination program-Haemophilus influenzae type b vaccine.

Haemophilus influenzae was first identified in 1892 by Pfeiffer, who wrongly deduced that the bacterium was the cause of influenza. The bacterium is a small (1- by 0.3-um) gram-negative organism of variable shape. The bacteria are often described as a pleomorphic coccobacillus. Six major serotypes of H. influenzae have been recognized (a-f); they are identified by antigenically distinct polysaccharide capsules. Strains lack a polysaccharide capsule and are referred to as nontypable strains. Type b and nontypable strains are the most relevant strains clinically, although encapsulated strains other than type b can cause disease.

The antigenically distinct type b capsule is a linear polymer made of ribosyl-ribitol phosphate. Strains of H. influenzae type b (Hib) cause disease primarily in infants and children under the age of six years. Nontypable strains are primarily mucosal pathogens, although the incidence of invasive disease caused by these strains is increasing. The most serious manifestation of infection with H. influenzae is meningitis. Haemophilus influenzae causes also epiglottitis (a life-threatening infection involving cellulitis of the epiglottis and supraglottic tissues), Cellulitis, and pneumonia in infants. Other less common invasive conditions can be important clinical manifestations of Hib infection in children include osteomyelitis, septic arthritis, pericarditis, orbital cellulitis, endophthalmitis, urinary tract infection, abscesses, and bacteremia without an identifiable

31

focus. Infections due to Haemophilus influenzae are uncommon among patients older than 6 years.

The most dependable method for making a diagnosis of Haemophilus influenzae infection is the isolation of the organism in culture. Therefore, the CSF of a patient in whom meningitis is suspected should be subjected to Gram's staining and culture. The presence of gram-negative coccobacilli in Gram-stained CSF is strong evidence for Hib meningitis. Recovery of the organism from CSF confirms the diagnosis.

Cultures of other normally sterile body fluids, such as blood, joint fluid, pleural fluid, pericardial fluid, and subdural effusion, are confirmatory in other infections. Detection of PRP is an important adjunct to culture in rapid diagnosis. Immunoelectrophoresis, latex agglutination, coagglutination, and enzyme-linked immunosorbent assay are effective in detecting PRP. These assays are particularly helpful when patients have received prior antimicrobial therapy and thus are especially likely to have negative cultures.

Childhood bacterial meningitis

Bacteria are still responsible for many cases of meningitis throughout the world, and bacterial meningitis remains one of the most potentially serious infections in infants and children with high risk of acute complications and chronic morbidity.

The mortality from these infections has dropped in recent years, but there is alarming evidence that the long term effects of meningitis early in life may be worse than previously thought. Reduced immunologic response to specific pathogens associated with young age is probably the most important risk factor for meningitis.

Approximately 95% of reported cases of meningitis occur between 1 month and 12 month of age. In many geographic areas the three most common bacterial pathogens causing meningitis are Haemophilus influenzae, streptococcus meningitis, and Niesseria meningitidis.

Haemophilus influenzae was the most common bacterial pathogens causing meningitis during the first year of life before the introduction of vaccines against Haemophilus influenzae type b (Hib).The wide spread use of Hib vaccines was associated with marked decrease in the frequency of Haemophilus infection, and Haemophilus influenzae is no longer the most common bacterial pathogens causing meningitis during the first year of life in many geographic areas such as UK and USA [2, 3, 4, 5.6].

The pattern of childhood bacterial meningitis in Iraqi children

In Iraq the precise and even the rough contribution of Haemophilus influenzae and other bacterial pathogens to acute bacterial meningitis and acute CNS infection in general remains unknown. It was not possible to find any published literature reporting the pattern of childhood bacterial meningitis. Many factors have been precluding the estimation of the contribution of Haemophilus influenzae and other bacterial pathogens to acute bacterial meningitis in Iraqi children. According to our experience at the University Hospital in Al Kadhimiyia (One of the three large teaching hospitals in Baghdad), the factors that have been precluding the determination the pattern of childhood bacterial meningitis in Iraqi children include:

1-The vast majority of our patients were receiving antibiotics for several days before referral and the CSF microscopic examination and CSF cultures didn't reveal the causative organism. Remembering that it has been shown that virtually all patients will have sterilization of the CSF within 24 hour of the initiation of antibiotics[5].In most instances the clinical suspicion of bacterial meningitis was supported by finding of CSF pleocytosis with or with out low CSF sugar or elevated proteins together with a satisfactory and relatively early response to antibiotics that cover the three common pathogens that are responsible for bacterial meningitis in many geographic areas of the world.

During the 1990s, ampicillin and chloramphenicol were the most commonly used antibiotics in Iraqi hospitals for the treatment of cases of suspected bacterial meningitis. However, during the previous 8 years, there has been an increased use of third generation cephalosporins especially cefotaxime in patients with diagnosis of bacterial meningitis.

2-Iraq had been in a continuous turmoil during more than two decades; there had been a continuous series of financial, social, and political crises. Successive wars and long blockade played an important role in the generation of these complex and continuous crises.

These crises were associated with poor recording systems in hospitals, deterioration in residents training program together with a marked deterioration in the academic and research performances in Iraqi hospitals compared with many other counties with much lower resources such as India and Pakistan.

The introduction of the Haemophilus influenzae vaccines into the developed counties vaccination program

The introduction of the Haemophilus influenzae type b (Hib) polysaccharide-conjugate vaccine into the developed counties (UK, and USA) vaccination program has had a dramatic effect. The incidence of Hib meningitis has decreased from around 2500 cases per year to less than 40 per year in some areas. Conjugate vaccines against H. influenzae type b (Hib) and N. meningitidis group C are now routinely given in the UK as part of the primary course of immunization at 2, 3 and 4 months.

An effective vaccine against N. meningitidis group B is not yet available. Trials of conjugate pneumococcal vaccines are underway and may be introduced in the UK in the near future [4]. Haemophilus influenzae was once the most common cause of bacterial meningitis in the United States. The incidence of H. influenzae meningitis declined precipitously following the introduction of the H. influenzae type b (Hib) vaccine in 1987, and H. influenzae now accounts for <10% of bacterial meningitis cases[7].

Hib vaccines have been shown to be safe and immunogenic during the firs month of life. The efficacy rate of Hib vaccines ranged from 70-100%.According to the committee on infectious diseases of the American academy of Pediatrics recommendation, all children should be immunized with Haemophilus influenzae type b (Hib) vaccine at about 2 month of age or as soon as possible later [8, 9, 10].

Rational for the possible introduction of Haemophilus influenzae vaccine into the Iraq vaccination program

Many of the new therapies and preventive measures including new antibiotics(e.g. third generation cephalosporins) and new vaccines(e.g. Hepatitis B vaccine) have been introduced in Iraq with a beneficial effect based on outside researches and experiences rather than Iraqi researches and clinical studies. Obviously, improvement in the specific diagnosis of bacterial meningitis can't be witnessed in Iraqi hospital in the near future. It seems logic that Hib vaccines should be introduced into the Iraq vaccination program depending on evidences available from outside Iraq.

References

1-Murphy TF. Haemophilus infections. In: Harrison's Principles of Internal Medicine 15th ed. CD-ROM.

2-Baraff LJ, Lee SI, Schringer DL. Outcomes of bacterial meningitis in children: a meta-analysis. Pediatr Infect Dis 1993; 12:393.

3-Pomeroy SL, Holmes SJ, Dodge PR, et al. Seizures and other neurologic squeale of bacterial meningitis in children. N Engl J Med 1990; 323:1651.

4-Taylor HG, Mills EL, Clampi A, et al. the squeale of Haemophilus influenzae meningitis in school-age children.N Engl J Med 1990;323:1657.

5-Hargreaves RM, Slack MP, Howard AJ. et al. Changing patterns of invasive Haemophilus influenzae disease in England and Wales after introduction of the Hib vaccination programme. BMJ 1996; 312:160.

6-Schuchat A et al: Bacterial meningitis in the United States in 1995. N Engl J Med 1997; 337:970.

7-Blazer S, Bernat M, Alon U.Bacterial meningitis. Effect of antibiotic treatment on the cerebrospinal fluid. J Clin Pathol 1983; 92:480.

8-Committee on Infectious Diseases: Haemophilus influenzae type b conjugate vaccines: Recommendations for immunization with recently and previously licensed vaccines. Pediatrics 1993; 92:480.

9-Committee on Infectious Diseases: Haemophilus influenzae infections, in 1997 Red Book, Report of the Committee on Infectious Diseases, G Peter et al (Eds). Elk Grove Village, IL, American Academy of Pediatrics, 1997.

10-Peltola H et al: Perspective: A five-country analysis of the impact of four different Haemophilus influenzae type b conjugates and vaccination strategies in Scandinavia. J Infect Dis 1999; 179:223.

Cardiac Resynchronization Therapy at Al-Kadhimiyia Teaching Hospital in Baghdad

The New Iraqi Journal of Medicine 2007; 3 (2): 9-15.

Rafid B.Hashim Al-Taweel
Al Kadhimiyia Teaching Hospital
Al Kadhimiyia Baghdad Iraq

Abstract

Background: About 30% of patients with chronic heart failure have evidence of a major intraventricular conduction delay, which may worsen the left ventricle (LV) systolic dysfunction through asynchronous ventricular contraction. Uncontrolled studies suggest that multi-site biventricular pacing improves hemodynamics and well being by reducing ventricular asynchrony. The aim of this paper was to assess the clinical efficiency of this therapy.

Patients and Methods: From November 2004 to November 2005, 22 patients (86.5% males, 13.5 females) with moderate to severe heart failure were observed at Al-Kadhimiyia Teaching Hospital. Their ages ranged from 30 to 77 years, (mean age 59.5 yr) mean 59.5 ± 2.5. All of them had conduction delays not responding to medical therapy. They remained symptomatic despite pharmacologic treatment with (ACEI, ARB, $B_.$-blockers, and diuretics with digoxin as needed. Therefore were subjected to Cardiac Resynchronization Therapy (CRT).

Successful implantation was achieved in about 19 patients (86.5%) and failed in 3 others (13.5%) due to technical difficulties, bifocal RV pacing was tried on one of later 3 patients. Follow up of patients who had successful CRT was possible only in 15 patients (79%); each was assessed within the first 3 months after CRT, while 4 patients (21%) couldn't be followed up.

Results: During an average 3 months of follow up there was a significant improvement in functional capacity, 6 minutes walk distance, ECG, and echocardiographic.

Conclusion: Although it was technically complex, atriobiventricular pacing significantly improved exercise tolerance, and functional capacity.

Introduction

Congestive heart failure remains a major and growing public health problem. Medical therapies by drugs such as Angiotensin Converting Enzyme Inhibitor (ACEI), Angiotensin Receptor Blocker (ARD), Beta blockers, sparinolactone) may not always be successful.

About 30% of patients with chronic heart failure have evidence of a major intra-ventricular conduction delay, which may worsen the left ventricle (LV) systolic dysfunction through asynchronous ventricular contraction. Uncontrolled studies suggest that multi-site biventricular pacing improves hemodynamics and well being by reducing ventricular asynchrony.

Cardiac resynchronization therapy (CRT) has been successfully used in patients with heart failure and ventricular conduction delays. It has been found that CRT restores synchronous activation of the LV, resynchronizes the timing of activation of (both the RV and LV), and optimizes the atrio-ventricular filling by reducing the AV delay, thus abolishing the ventriculo-atrial gradient and minimizing the presystolic mitral regurgitation (MR).These effects result in major improvements in stroke volume, ventricular contractility, reduction of diastolic and systolic mitral insufficiency. Pacing the left ventricle is performed with a specially designed trans-venous leads inserted into a distal cardiac vein through the coronary sinus to pace the LV free wall.

Bifocal RV has been used to overcome the difficulties faced when cannulating the coronary sinus because of anatomical variabilities and studies are in progress to evaluate it's efficacy in resynchronization to produce narrower QRS on ECG and better hemodynamics results (increase in Ejection fraction-EF, cardiac output and reduction in MR).

Bifocal RV endocardial stimulation can be achieved through the presence of 2 leads in the RV, one in the RV apex and the other high in the interventricular septum in its weak fibrous portion which upon stimulation act as if it's stimulating the LV. This approach is developed in order[1,2,3,4,5,6,7,8,9,10].

Patients and Methods

From November 2004 to November 2005, 22 patients (86.5% males, 13.5 females) with moderate to severe heart failure were observed at Al-Kadhimiyia Teaching Hospital.

Their ages ranged from 30 to 77 years, (mean age 59.5 yr) mean 59.5 ± 2.5 .17 patients(78%) had ischemic Cardiomyopathy (CMP) (post myocardial infarction),three patients(13%) had idiopathic dilated CMP, and two (9%) had tachyarrhythmia). Figure (1) shows the etiology of heart failure in this study. All of them had conduction delays not responding to medical therapy. They remained symptomatic despite pharmacologic treatment with (ACEI, ARB, B.-blockers, diuretics with digoxin as needed. and therefore were subjected to Cardiac Resynchronization Therapy (CRT).

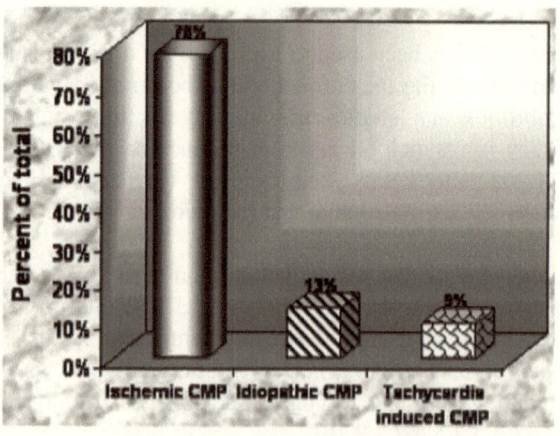

Figure (1): The etiology of heart failure in this study

Successful implantation was achieved in about 19 patients (86.5%) and failed in 3 patients (13.5%) due to technical difficulties (The anatomical variations in coronary sinus and the cardiac veins from patient to patient causes considerable difficulties during implantation and lead to insertion through the coronary sinus, hence failure for which bifocal RV pacing was tried in one of these patients as we'll mention later. They were assessed for the improvement in functional capacity, exercise tolerance, electrocardiographic parameters, and echocardiography indicators.

Echocardiographic indicators were compared before and after CRT in about 3 months period in same patients.
The functional capacity (NYHA class), 6- min. walk distance, electrocardiographic findings (QRS duration) and echocardiographic indicators (MR and its severity, and EF) were assessed before and after the implantation of the device (three months after CRT). The available dyssynchrony marker used in the study was ECG and the presence, severity of MR discovered by echo study was divided into mild, moderate and severe depending on the quantitative measure of the jet flow of MR by color Doppler study. Mild =blood jet just traversing the mitral valve. Moderate =blood jet reaching the mid atrium. Severe =blood jet reaching the left atrial wall.

Follow up of patients who had successful CRT was possible only in 15 patients (79%), each assessed within the first 3 months after CRT, while 4 patients (21%) lost follow up from the beginning, so the parameters compared in this study were applied only on 15 patients whom we followed regularly.

Results

Table-1 summarizes the effects of CRT on functional capacity, exercise tolerance, electrocardiographic parameters, and echocardiographic indicators. Figure-2 shows the Six minute walk distance before and after CRT.

Figure-3 (A and B) shows QRS duration before and after CRT. Figure (4) shows the ejection fraction on echo study before and after CRT. Figure (5) shows the severity of MR before and after CRT.

Limitations and complications

2 out of 15 patients (13%) experienced secondary failure of CRT after a successful implantation with improvement in clinical outcome and echocardiographic indicators. One patient (6.5%) complained from phrenic nerve stimulation perceived by the patient as a pricking feeling with each pacing.

	Before CRT	3 months after CRT
Functional capacity (NYHA class)	80% of patients were in grade III while 20% of patients were in grade IV	Nearly100% were in grade II in all the patients (P<0.001)
The mean 6-min.-walk distance (exercise tolerance)	137.2 m.±7.35 SEM.	increased up to 213.4 m. ±10 SEM (P<0.001) i.e. there was an increment of about 76.2 m. (55.5%)
ECG finding	2 patients (13%) had narrow QRS (≤120msc.) before CRT, while the remaining patients had wide QRS in the form of LBBB (87%). Mean pre CRT QRS duration in all patients was 176.3msc.± 10.2 SEM	QRS duration in 14 patients (93%) was within 120 msc.In 1 patient (7%) was about 140 msc. with a mean value in all of them of 121.3 msc.±1.7 SEM (P<0.001) PR interval was set at about 120 msc in all patients.
Echocardiographic indicators	Nearly all of them had EF≤40% with a mean value of 36.47%± 1.1 SEM	Increased to 44.93±1.2 SEM (P<0.001) i.e. 8.5% increase in EF
MR	Mild in 27%, moderate in 66%, severe in 7%	About 80% had mild MR and the rest didn't have MR (P<0.01)

Table-1: The effects of CRT

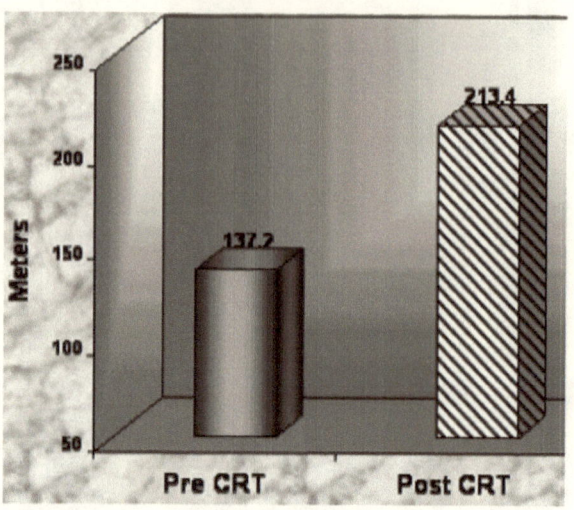

Figure (2): The Six minute walk distance before and after CRT

Echocardiographic indicators

As they had reduction in EF to less than 40%, NYHA class became grade III-IV, reduction in 6-minute walk distance with a wide QRS, and this was mostly due to LV lead dislodgement. Also one patient (6.5 %) with post- MI died during follow up after progressive deterioration in his condition although he had an initial improvement.

Bifocal RV pacing was tried for the first time in one of the patients who had primary failure of the biventricular CRT, and was followed up within one month, there was:

Improvement in NYHA class from grade III to grade II (P< 0.001).Improvement in 6 –min- walk distance from 120 m. to 180 m. i.e. 60 m. (50%) increase in distance (P <0.001) Reduction in QRS duration from 180 msc. to 120 msc. (P<0.001).Improvement in LVEF was from 35% to 42% (P< 0.01).Reduction in severity of MR from moderate to mild (P<0.01).

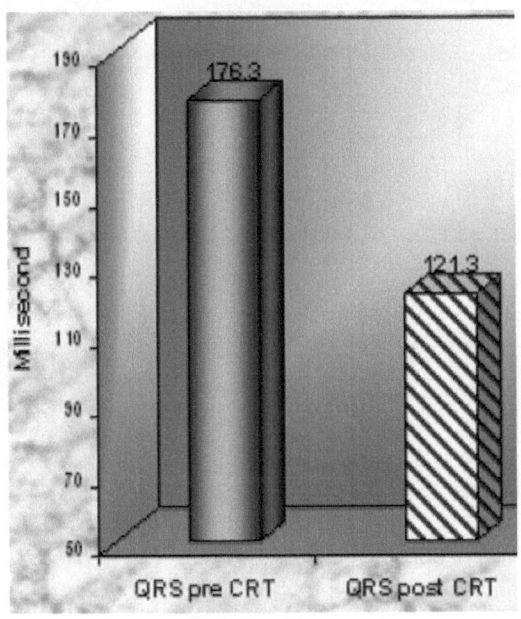

Figure-3 (A): QRS duration before and after CRT.

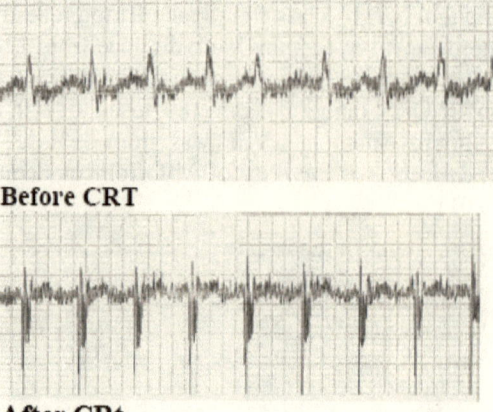

Before CRT

After CRt

Figure-3 (B): QRS duration before and after CRT.

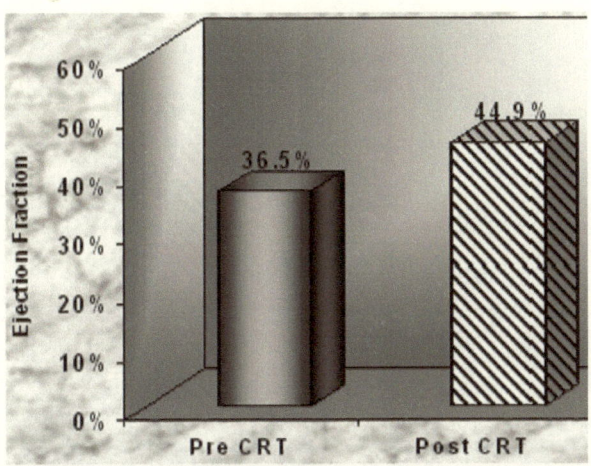

Figure (4): The ejection fraction on echo study before and after CRT.

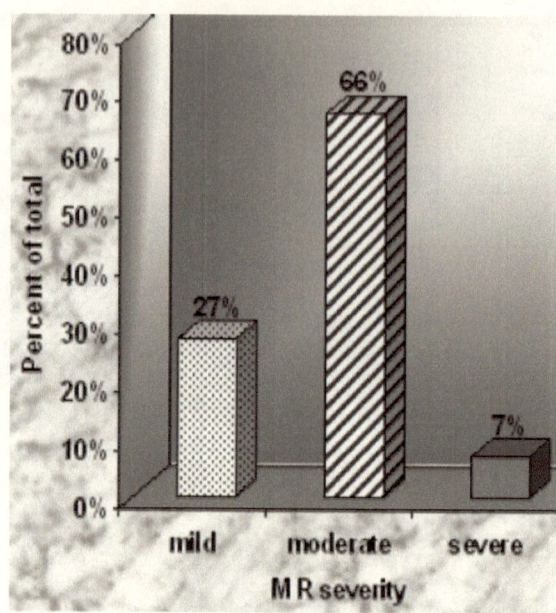

Figure.5 (A): The severity of MR before CRT.

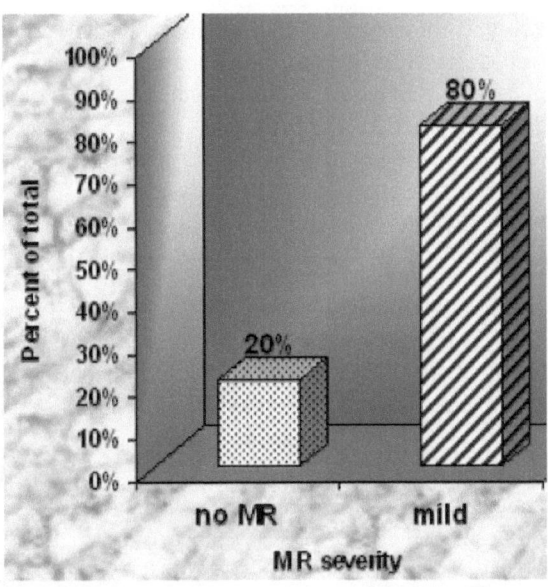

Figure.5 (B): The severity of MR after CRT.
Discussion

In this study, the success rate of implantation (86.5%) was comparable to previous reports (88%-92%).The improvement in functional capacity, was represented 6 min walk distance, ECG, and echocardiographic after CRT was also similar to previous studies. The complication of secondary failure of CRT with subsequent deterioration in clinical state and echocardiographic indicators were most likely due to LV Lead dislodgment has also been reported before. Fortunately, coronary sinus perforation was not encountered in any of our patients, but phrenic nerve stimulation (i.e. diaphragmatic contraction) was present in only one patient, probably because we could not achieve good resynchronization without maximizing the energy of LV lead output (7.5 mv) at the expense of this annoying complication, however this complication remains low[11,12,13,14,15,16,17,18,19,20].

Acknowledgement

Dr Rafid B. Hashim Al-Taweel conducted this study when he was a doctoral student at Al Kadhimiyia Teaching Hospital. This paper represents his doctoral dissertation. The CRT therapy was done by Dr Ammar Hamdi (a consultant internist) and the supervisor of Dr Rafid's study.

References

1-Leclercq C, Kass DA. Retiming the failing heart: principles and current clinical status of cardiac resynchronization J Am coll Cardiol. 2002; 39:194-201.

2-Nelson GS, Berger RD, Fetics BJ, Talbot M, Spinelli JC, Hare JM, Kass DA. Left ventricular or biventricular pacing improves function at diminished energy cost in patients with dilated CMP and LBBB. Circulation 2000; 102:3053-9.

3-Blancc JJ , Etienne Y , Gilard M , Mansourati J , Munier S , Boschat J, Benditt DG , Lurie KG. Evaluation of different ventricular pacing site in patients with severe heart failure. Results of an acute haemodynamic study. Circulation. 1997; 96:3273-3277.

4-AbrahamWT, Hayes DL. CRT for heart failure. Clinical studies of CRT. Circulation 2003; 108(21): 2596.

5-Mateos JC, Alboronz RN, Mateos EI, Gimenez VM, Pachon MZ. Right Ventricular Bifocal in the Treatment of Dilated Cardiomyopathy with Heart Failure. Results. Arq. Bras. Cardiol. vol.73 n.6 Sao Paulo. 1999.

6-Abraham WT, Hayes DL. CRT for heart failure. Clinical implications of CRT. Circulation 2003; 108(21): 2596.

7-Abraham WT, Fisher WG, Smith AL et al. Multicenter Randomized Clinical Evaluation. Cardiac resynchronization in chronic heart failure N Engl J Med 2002, 346:1845- 53.

8-Linde C, Leclerq C, and Rex S, et al. Long term benefits of biventricular pacing in congestive heart failure: Results from the Multisided Stimulations in Cardiomyopathies (MUSTIC) study. J Am Coll Cardiol 2002; 40:111-8.

Prevalence of post traumatic stress disorder in primary school children in Baghdad

The New Iraqi Journal of Medicine 2007; 3 (2): 16-19.

AliH.Razoki
Consultant psychiatrist
Yarmok teaching hospital Baghdad
Ghazi Aboud
Consultant psychiatrist
Al Rashad Hospital, Baghdad
Khawla A.Al-qaisy
Director of the centre of educational and psychological research

Abstract

Background: The present situation in Iraq dominated by violence, looting, kidnapping, torture, or murder has created an extremely threatening and traumatizing atmosphere for the whole population especially children.

Objectives: to determine the prevalence of post traumatic stress disorder (PTSD) among primary school children in Baghdad.

Method: A cross sectional multi -stages sample survey of 979 respondent aged 9-15 years had been contacted in Baghdad during the period from January to March.2005, six schools were involved with 150-160 children representing each school.

Tool: Arabic version of MINI (international neuro-psychiatric interview PTSD module 1 was applied.

Results: During the last 2 years, 58% of the respondents had experienced major traumatic event .The prevalence of PTSD among schoolchildren was 18%. The male/female ratio was 2/3, which was statistically significant.

Conclusion: The traumatizing events were very common and had precipitated PTSD in 18% of children. The prevalence was less than expected which may suggest that Iraqis had adapted some sort of psychological immunization during the last three successive stressful & traumatizing decades. The study warrants exceptional efforts to re-stabilize the situation in order to avoid wide spread of morbidity and disability among children.

Introduction

The critical current situation and the tragic amount of violence in Baghdad have affected the whole population causing different degrees of distress, anxiety, fear and depression. People who had been exposed directly to severe traumatic events could experience severe stress especially children. Psychological trauma may precipitate post traumatic stress disorder (PTSD) in some of them resulting in along time disturbance and even disability to the affected children. We hope that this study will participate in estimating of the magnitude of the problem and to help planning to deal with it in the future.

The prevalence rate of PTSD was 20.4% in Afghanistan [1] .PTSD is common in a large representative sample: In USA. Kessler et al estimated the lifetime prevalence of PTSD to be 7.8% (women 10.4% men 5%). PTSD can affect anyone during his life and up to 30% of people susceptible to stressful events or state of unexceptionally threatening or catastrophic nature (such as a natural disaster, war, torture, rape, sexual abuse) will go on to develop to PTSD [2] According to DSM IV, PTSD is an abnormal emotional reaction to traumatic event which is characterized by symptoms of re-experience, hypervigilance, and avoidance. The aim of this study is to determine prevalence of PTSD in primary school children in Baghdad.

Patients and methods

A cross sectional study (Prevalence survey) was conducted in 6 schools in Baghdad. Multi-stages sampling method was used involving the 2 educational Sectors of Baghdad (Karkh and Risafa) as the base for stratification with a size of 979. In every stratum, three schools were chosen randomly forming a sum of six schools. In each school six classes was randomly chosen .Using systematic random sampling 25 children from each class was enrolled. The grades 4, 5, 6 had been involved only. Arabic version of M.I.N.I (International Neuro-psychiatric Interview PTSD module 'I') was applied (David Sheehan, University of South Florida.

Data collection techniques: two well-trained psychologists conducted direct interviews with the children.

Statistical analysis: Descriptive statistics for the prevalence and demographic features, and correlation statistics for associated features using EPI and SPSS11 soft ware was used.

Results

979 primary school children were interviewed (492 male, 487 female), from six different schools in Baghdad taking 150-160 children from each school and 50-55 from grades 4, 5, 6. Their ages ranged from 9-15 years (with the average 11 years) .Figure (1) shows the age and sex distribution of enrolled children. 578 (58%) children were exposed to a major traumatic event during the last two years. 177 (71 males 40% and 106 females 60%) were affected by PTSD giving a prevalence rate of 18%. Male to female ratio was 2\3 using test of proportion Z=-2.564 which is more than 0.96. P value is less than 0.05 so this ratio is statistically significant. Chi squire was 19.95 with 8 degrees of freedom .P value was 0.01.

Avoidance symptoms were as such: 85% avoided thinking about the event, 77%had trouble in recalling some important part of what happened, 64% became less interested in hobbies and social activities, 49% felt detached and estranged from others, 54% noticed their feelings are numbed, and 55% felt their life would be shortened. Hyper arousal symptoms were as such: 68% had sleeping difficulty, 50% were irritable, 69% had difficulty in concentrating, 43% were nervous, and 61% were easily startled. The symptoms significantly interfered with the schoolwork, social activities, or caused significant distress in 83% of the sample.

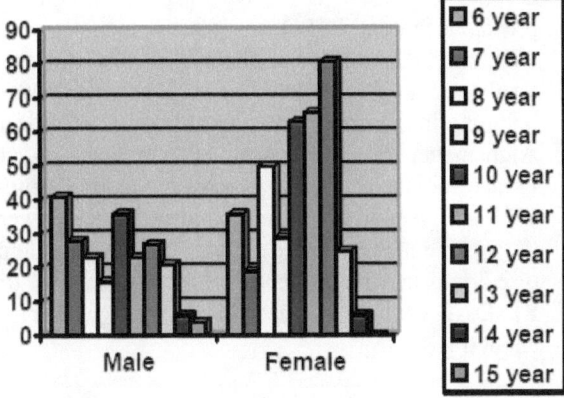

Figure (1): The age and sex distribution of enrolled children

Discussion

58% of the sample of children had been exposed to a major traumatic event during the last 2 years, which points out to the volatile and violent environment we are living in and especially our children. This number indicates the violent traumatic atmosphere, which needs exceptional efforts to re-stabilize the situation in order to avoid widespread morbidity and disability amongst those children.

18% of the children met the diagnostic criteria of PTSD, which may suggest that Iraqis and especially children had been adapted in one way or another to stressful and traumatic events probably because these events became chronic and acted as a psychological immunization. The prevalence rate of PTSD was lower than expected compared with similar studies elsewhere[2] 22.2% African schools Seedat et al[3] ,while it was 35% after US embassy bombing in Nairoby Njenga et al [4]; 41% Gardoza et al in Afghanistan [5] ,and it was 31% in bombing victims in France Verger et al [6]. However, the prevalence rate of 18% is high enough to cause concern to take preventive and therapeutic measures to deal with it. The male \female ratio of 2/3 indicates that females are more vulnerable to develop PTSD. These results open the door for further researches about this subject.

Conclusion

The impact of the current extremely violent environment in Baghdad on children is significant. There is a need for an effective plan and a program to deal with the psychological trauma and devastating effect. Further researches in this field are essential.

References

1-Scholte WF, Olff M, Ventevogl P, Vries GJ , Jansveld E, Cardozo BL , Crawford CAG. Mental health symptoms following war and repression in eastern Afghanistan. 2004 Amer JAMA Aug. 2004; 292 (5).

2-www.Prevalence and incidence of post-traumatic stress disorder-wrong diagnosis.com.

3-Seedat S, Nymal C, Njenga F., Vythilingum B, Stein DJ. Trauma exposure and PTSD symptoms in urban African schools. British journal of psychiatry 2004, 184, 169-175.

4-Njenja FG, Nicholls PJ, Nyamai C, Kijamwa P, Davidson JR. post traumatic stress after terrorist attack: psychological reaction following the US embassy bombing in Nairooby, British J.Psych 2004); 185,328-333.

5-Cardozo Brbara Lopes, Bilukha OO, Crawford Carol AG, etal. Mental health, social functioning, and disability in postwar Afghanistan, JAM 2004; 292 (5).

6-Verger P, Dab W, Lamping DL, et al. The Psychological Impact of Terrorism: An Epidemiologic Study of Posttraumatic Stress Disorder and Associated Factors in Victims of the 1995-1996 Bombings in France. American J Psych 2004; 1384-1389.

Antithymocyte globulin with cyclosporine A and methylprednisolone in the treatment of aplastic anemia adult patients: Iraqi experience

The New Iraqi Journal of Medicine 2007; 3 (2):24-27.

Abdul Majeed A.
Hammadi, Alaa Makki ,
Amel A. Ali , Waleed A. Azeez
Bone marrow transplantation section. Baghdad Medical City

Abstract

Background: Antithymocyte globulin (**ATG**) has been the standard alternative for hematopoietic stem cell transplantation (HSCT) or immunosuppressive therapy (IST) of aplastic anemia. The aim of this study was to describe our experience with **ATG** in the treatment of aplastic anemia.

Patients and methods: From October 2003 to March 2007, 30 Iraqi patients (16 males, and 14 females) 24 with severe aplastic anemia (SAA) (=80%) and less severe aplastic anemia (nSAA) (n = 20%) patients were assigned to treatment with rabbit **ATG** (Thymoglobulin; Fresenius Germany) and methylprednisolone. Their ages ranged from 14-55 years. The patients were treated with 1-ATG 10-15mg/kg 5 –7 days, 2-cyclosporine 5mg/kg 6 months,3-methyl prednisolone 5mg/kg oral or iv 5 days and taper in 28 days.

Results: 12 (40%) patients achieved partial remission (P. Overall response was 50%. 18 remained alive. 4 patients (15%) didn't show any response. Six patients (20%) died within 12 months of starting treatment, 4(13.3%) died after more than 12 months after starting treatment.

Conclusion: Our experience is similar to the previously reported experiences [11] with similar overall response to the previously reported response of 40-70%, and similar problems.

Introduction

Aplastic anemia is thought to be an immune-mediated bone marrow disease, characterized by bone marrow aplasia and peripheral blood pancytopenia. Most patients can be successfully treated with either hematopoietic stem cell transplantation (HSCT) or immunosuppressive therapy (IST) and can survive long term [1]. Because HSCT cures aplastic anemia, it is the treatment of choice for young patients with suitable stem cell donors. HLA-matched related donors are widely accepted as stem cell donors [2] whereas unrelated donor HSCT still carries a significant risk for morbidity and mortality [3]. Most patients do not have donors for standard-risk HSCT and rely on IST as the first-line treatment. and in our country stem cell transplant is not feasible. Antithymocyte globulin (ATG) has been the standard for IST of aplastic anemia. ATG significantly improves survival compared with supportive care or androgen therapy. Response rates vary between 40% and 70%, and long term survival after ATG-based IST is similar to that in unselected patients treated with HSCT [3, 4, 5, 6, 7].

Patients and methods

From October 2003 to March 2007, 30 Iraqi patients (16 males, and 14 females) 24 with severe aplastic anemia (SAA) (=80%) and less severe aplastic anemia (nSAA) (n = 20%) patients were assigned to treatment with rabbit ATG (Thymoglobulin; Fresenius Germany) and methylprednisolone. Their ages ranged from 14-55 years. The patients were treated with 1-ATG 10-15mg/kg 5 –7 days, 2-cyclosporine 5mg/kg 6 months, 3-methyl prednisolone 5mg/kg oral or IV 5 days and taper in 28 days see attachment 1. Patients were assessed prior and post treatment. Assessment included complete blood counts, transfusion history, medication, illnesses or adverse events, and general health status, full history and clinical examination, checking for osteoporosis/necrosis, renal failure, hypertension, clinical or laboratory evidence for paroxysmal nocturnal hemoglobinuria (PNH) (Ham test, haptoglobin),side effects of treatment as recorded in their case sheets myelodysplastic syndrome (MDS), leukemia, and solid tumors.

Response was defined as a significant improvement of blood counts (partial or complete remission) within 4 months according to the current consensus guidelines of the European Blood and Marrow Transplant (EBMT) Severe Aplastic Anemia Working Party [7].Complete remission (CR) was defined as blood counts normal for age and sex. For patients 15 years and older, we used hemoglobin (Hb) levels 12 g/dL or greater for women and Hb levels 13 g/dL or greater for men, granulocyte counts 1.5×10^9/L or higher, and platelet counts 150×10^9/L or higher. For 6 children, age-adjusted lower thresholds were used [8].

Partial remission (PR) was defined by transfusion independence and by an unsupported increase of the counts of at least one cell line over baseline values (Hb by at least 3 g/dL, granulocytes by at least 0.5×10^9/L, if previously lower than 0.5×10^9/L, and platelets by at least 20×10^9/L, if previously lower than 20×10^9/L) or by doubling or normalization of counts of at least one cell line if previous counts of the respective cell line(s) did not meet the criteria for SAA. All remissions had to be confirmed by at least 2 blood counts at least 4 weeks apart.

Clinical relapse was defined as a decrease in any of the peripheral blood counts to less than 50% of the median sustained counts during remission, by return of the counts to levels meeting the definition of SAA, or by the need for transfusion.

The patients were followed for more than 3 years that includes full clinical assessment and complete blood counts in addition to any clonal evolution including (PNH acute leukemia) and following type of response and cyclosporine dependence Updated information was obtained from all patients. The median observation time of surviving patients is 3year.

Results

About four months after the start of treatment, 13 patients (43.1%) achieved at least PR and were considered responders. 2 patients (6.6%) achieved CR .At the end of the follow-up, 12(40%) patients achieved PR.

The overall response was 50%. The patients who are alive are 18 out of 30 cases. 4 patients (15%) didn't show any response. Six patients (20%) died within 12 months of starting treatment, 4 patients (13.3%) died after more than 12 months after starting treatment. Causes of death were considered cerebral hemorrhage in 2 patients, respiratory failure from pneumonia in 3 patients, multi-organ failure from sepsis in 4 patients, and sepsis and leukemia in one patient.

Blood counts of 13 patients improved more than 3 months and up to 2 years after treatment. Three of these patients reached CR and continue to be in stable condition need no additional courses of IST. A late remission was not seen in our cases.

Of all patients who had a response to a regimen containing CsA (4 of 30) required administration of the drug for more than 6 months because their counts decreased with the discontinuation of CsA or when the dose of the drug was lowered, and they returned to previous counts with the re-administration of CsA.

Side effects: Osteonecrosis occurred in one patient, hypertrichosis occurred in 4 female patients; gingival hyperplasia occurred in 7 patients, hypertension requiring treatment occurred in 2 patients, and muscle cramps occurred in 6 patients.

Discussion

Approximately two thirds of the patients with aplastic anemia respond to a combination of ATG, CsA, and short-course methylprednisolone. This combination was significantly more effective than ATG and methylprednisolone alone (response rate, 70% versus 41%; $P = .015$). The advantage of adding CsA was only evident in patients with SAA, Favorable early results of the treatment with ATG, methylprednisolone, and CsA have been confirmed in several small series and in large American [9], and Italian trials [10]. CsA alone has activity in aplastic anemia, but European trials showed that the combination of ATG and CsA is more effective than CsA alone [12]. Aplastic anemia may be particularly sensitive to CsA in children [13]. Combined ATG, methylprednisolone, and CsA have thus become the standard immunosuppressive protocol in children and adults with SAA against which new treatment regimens are compared.

Conclusion

Our experience is similar to the previously reported experiences11] with similar overall response to the previously reported response of 40-70%.

References

1-Young NS. Acquired aplastic anemia. Ann Intern Med. 2002; 136:534-546.

2-Deeg HJ, Leisenring W, Storb R, et al. Long-term outcome after marrow transplantation for severe aplastic anemia. Blood. 1998; 91:3637-3645.

3-Deeg HJ, Seidel K, Casper J, et al. Marrow transplantation from unrelated donors for patients with severe aplastic anemia who have failed immunosuppressive therapy. Biol Blood Marrow Transplant. 1999; 5:243-252.

4-Champlin R, Ho W, Bayever E, et al. Treatment of aplastic anemia: results with bone marrow transplantation, antithymocyte globulin, and a monoclonal anti-T cell antibody. Prog Clin Biol Res. 1984; 148:227-238.

5-Camitta B, O'Reilly RJ, Sensenbrenner L, et al. Antithoracic duct lymphocyte globulin therapy of severe aplastic anemia. Blood. 1983; 62:883-888.

6-Bacigalupo A, Brand R, Obeto R, et al. Treatment of acquired aplastic anemia: bone marrow transplantation compared with immunosuppressive therapy the European Group for Blood and Marrow Transplantation experience. Semin Hematol. 2000; 37:69-80.

7-Schrezenmeier H. Treatment of aplastic anemia with immunosuppression and hematopoietic growth factors: 25th Annual EBMT Meeting Educational Book. Hamburg, Germany: 1999; 123-131.

8-Lee GR, Foerster J, Lukens JN, Paraskevas F, Rodgers GM. Wintrobe's Clinical Hematology. 10th ed. Baltimore, MD: Lippincott Williams & Wilkins; 1998.

9-Rosenfeld SJ, Kimball J, Vining D, Young NS. Intensive immunosuppression with antithymocyte globulin and cyclosporine as treatment for severe acquired aplastic anemia. Blood. 1995; 85:3058-3065.

10-Bacigalupo A, Bruno B, Saracco P, et al. Antilymphocyte globulin, cyclosporine, prednisolone, and granulocyte colony-stimulating factor for severe aplastic anemia: an update of the GITMO/EBMT study on 100 patients. European Group for Blood and Marrow Transplantation (EBMT) Working Party on Severe Aplastic Anemia and the Gruppo Italiano Trapianti di Midolio Osseo (GITMO). Blood. 2000; 95:1931-1934.

11-Raghavachar A, Kolbe K, Höffken K, et al. Standard immunosuppression is superior to cyclosporine/filgrastim in severe aplastic anemia: the German multicenter study [abstract]. Bone Marrow Transplant. 1999; 23(suppl 1):S31.

The role of percutaneous pure spirit injection in treatment of benign, solitary, cold, cystic nodule of the thyroid gland

The New Iraqi Journal of Medicine 2007; 3 (2):28-31.

Hassan Al-Sikafi
Consultant surgeon
Baghdad Teaching Hospital
Malik Al Hashimi
Consultant surgeon
Baghdad Teaching Hospital

Abstract

Background: The role of percutaneous pure spirit injection of solitary thyroid nodules has been suggested by liviraghi in 1990 as a possible therapy for autonomously functioning toxic thyroid nodules Percutaneous 95% spirit injection has been considered a safe, low cost and effective therapeutic procedure in patient with benign cystic thyroid nodule. The aim of this paper is to report our experience.

Patients and method: During the years 2001 and 2002, 152 patients with clinically detected solitary thyroid nodules were observed at Baghdad Medical City, including 72 patients having single, cystic and cold nodules. Fine needle aspiration cytology(F.N.A.C)was done for all patients(72),57 patients were benign,6 patients were malignant or suspicious and 9 patients with indeterminate aspirate.60 patients were enrolled in a clinical study aiming at investigating the possible beneficial effects of pure spirit(ethanol 95%) injection.

Results: Pure spirit (ethanol) injection into thyroid nodule was done in only 60 patients.42 patients (70%) with benign cystic thyroid nodules were successfully treated with aspiration and pure spirit injection, of them 24 patients (40%) had complete disappearance of nodule, 18 patients(30%) had reduction in the size of nodule in more than (50%) versus baseline. Over 1-2 years period of follow up there was 2 recurrent cases (3, 3%).In 18 patients (30%) no significant reduction was obtained after month of spirit injection.

Conclusion: Percutaneous 95% spirit injection of cystic thyroid nodule was beneficial in 70% of cases suggesting a potential role for this therapy.

Introduction

Single thyroid nodule is a common thyroid problem. The large majority of these nodules are benign and the incidence of malignancy in cold solitary thyroid nodule is 10% 20%. [1,2].Ultrasound can provide information regarding the consistency of cold thyroid nodule being cystic or solid ,2-9% of cystic thyroid nodule may harbor malignant tumor[3,4].Fine needle aspiration cytology(FNAC) is safe, and may be used to diagnose, treat and sometimes cure the cystic thyroid nodule[5,6].The role of percutaneous pure spirit injection of solitary thyroid nodules was first proposed by liviraghi in 1990 as a possible therapy for autonomously functioning toxic thyroid nodules Percutaneous 95% spirit injection has been considered a safe, low cost and effective therapeutic procedure in patient with benign cystic thyroid nodule [7, 8].

Patients and methods

During the years 2001 and 2002, 152 patients with solitary thyroid nodules detected clinically and confirmed by ultrasound, and Tm thyroid scan were observed at Baghdad Medical City, including 72 patients (no females, and males) having single, cystic and cold nodules. Fine needle aspiration cytology (F.N.A.C) was done for all patients (72), 57 patients were benign, 6 patients were malignant or suspicious and 9 patients with indeterminate aspirate.

60 patients were enrolled in a clinical study aiming at investigating the possible beneficial effects of pure spirit (ethanol 95%) injection. Serum thyroid hormones were available for only 31 patients.60 patients were enrolled in a clinical study aiming at investigating the possible beneficial effects of pure spirit (ethanol 95%) injection. Pure spirit (ethanol) injection into thyroid nodule was done in only 60 patients as described by Lowhagen and others [6]. 42 patients (70%) with benign cystic thyroid nodules were successfully treated with aspiration and pure spirit injection

Technique: Standard technique was used as described by Lowhagen, FNAC was performed with patient in supine position and neck was extended, the patient was instructed to refine from swallowing. The nodule steadied by left hand, no. 21 gauge hypodermic needle attached to a20 cc syringe is inserted into with continuous suction in and out. When aspiration had been completed small amount (1-5ml) of 95% spirit is injected slowly on the bases of the aspirated volume without removing the needle [10, 11].

There was no need for bed rest or hospitalization for observation. Spirit injection was done once in 44 patients, twice in10 patients. And three times in 6 patients.

Follow up: after pure spirit injection, follow-up of the patients clinically and by ultrasound was done for 1-2 years. First visit was after 1 week to check the result of fine needle aspiration cytology and the size of nodule, second visit 1months later on, and then 6-12 months.

Result

Percutaneous 95% spirit (ethanol) injection of cystic thyroid nodule was done for 60 patients. In 12 patients spirit injection was not done because of small amount of material aspirated (0.1-0.2ml), of them 10 cases operated upon because of cytological findings were malignant, suspicious or indeterminate. After spirit injection we followed the patients clinically and by ultrasound scan for 1-2 years. One month after the treatment, nodule size reduction greater than 50% versus baseline was observed in 42 patients(70%) including 24 patients (40%) had complete disappearance of nodule, 2 cases recurrent in 6-12 months later on. In 18 patients (30%) no significant reduction in size of nodule (either no reduction or less than 50% versus baseline) Five of them operated upon because of cytological indication (malignant, suspicious or indeterminate)   or repeated rapid re-accumulation, the other 13 patients not operated on because they had benign cytological diagnosis, we kept them on close follow up for 1-2 years as shown in table2. Repeated aspiration and spirit injection (two or three times) was done in 16 patients, either because of re-accumulation or no reduction had been obtained. Complications requiring hospitalization were not observed.

73

Discussion

Verda and his colleagues have found that 80% of patients with benign cystic cold thyroid nodule gets nodular volume reduction greater than 50% after one month from pure spirit injection with 3% recurrence after 1 year of follow up [2].Papini et al have found non-toxic solitary thyroid nodules are successfully treated in 90-100% of patients by percutaneous ethanol therapy (PET) and recurrence rate in only 2-5% [4]. Panunizi et al [8] treated 30 patients with autonomous thyroid nodules by PET, his experience confirm an excellent response and symptoms of hyperthyroidism and hormonal level because normal and at ultrasound evaluation all nodules had significant shrinkage. In this study the overall end result over long term follow up was 40 patients (66.6%)had significant volume reduction of benign cystic thyroid nodule ,of them 24 patients (40%) had complete disappearance of nodules is not mentioned in the other studies. Thyroid damage induced by spirit characterized by coagulative necrosis and hemorrhagic infarction due to vascular thrombosis and is well defined from surrounding thyroid parenchyma, and in cystic nodules the spirit will irritate the wall of the cyst to induce fibrosis

Conclusion

Percutaneous 95% spirit injection of cystic thyroid nodule was beneficial in 70% of cases suggesting a potential role for this therapy.

References

1-DewanSS: incidence of thyroid cancer in solitary and multi-nodular goiter Bompany, Radiation Medicine Center, 1986, 15-23

2-Rendall-CH (FNA of thyroid nodules for 3 years) J.clinpathol, 1989; 42(1):23-7

3-SimeoneJE, Daniels GH, Mueller Pr (High resolution real time sonography of thyroid).Radiology; 1982; 145; 431-433.

4-Papini-E, pacella-CM; Verda G. Percutaneous ethanol injection of benign thyroid nodule. Section of Endocrinology, Opedale Regina. A postolorum, Albono Laziale, Rome, ItalyCli.Endorinol.Oxf.1995Apr; 5(2):147-50.

5-Ascaft MW. Management of thyroid nodules for three years J.Clinpath.1989; 42; 23-27.

6-H Balty and M.love.Short practice of surgery. Twentieth Edition H.K Lewis and Co.Ltd. London, 1988.Chapter37, p666, 671-672.

7-Liviraghi-Narid-F.Percutaeous ethanol injection in treatment of toxic thyroid nodule. Department of Internal Medicine Regina Apoatolorum hospital Albano, Italy Clin. Endocrinal. Oxf 1993; 76(2); 911-6.

8-Verda-G, papini-F, Gallotti-Q, et al. Percutaneous ethanol injection in the treatment of the cystic thyroid nodules. Italy.Clin. Endocrinal Oxf.1994; 41(6); 719-724.

Continuing Medical education: Principles, concepts, and standards

The New Iraqi Journal of Medicine 2007; 3 (2): 32-35.

Aamir Jalal Al Mosawi
Head department of pediatrics
University Hospital in Al Kadhimiyia
Al Kadhimiyia Baghdad Iraq

Abstract

Medicine is witnessing an incessant and great progress in all fields. The tremendous advances in the understanding of the bases of diseases have allowed more rational bases for the diagnosis and management of known medical disorders. Additional diagnostic tools and new therapies are continuously emerging and contributing to improved patient care and management, and hoist the expectation for more specific and curative therapy for many disorders. Therefore, the medical practice is rapidly changing. The aim of this paper is to briefly review the principles, concepts, and the generally accepted standards of continuing medical education and its impact on patient care.

Principles and concepts

The need of any physician to be always well informed on the modern methods in the diagnosis, recent treatment, prescribing, and rehabilitation of patients having a variety of diseases is universally accepted, but not universally put into practice. The need for Continuing Medical Education (CME) has become more complex because of the rapid advancements in medical sciences and technology.

Physicians may not be up to date in the knowledge and skills that were acquired during their college education and may therefore become professionally and functionally outmoded and less effective.

Therefore, there is need to recuperate physicians' skills and knowledge to keep them capable of efficiently leading the provision of health care. Obviously, personnel who did not have sufficient background education and are unlikely to benefit from rehabilitation programs focusing in training in newer technologies.

Physicians affiliated to university faculties and departments and are responsible for educating and training the upcoming health care personnel, are also expected to require rebuilding and retrieving their skills that may have become archaic. This also applies to physicians working in the health research institutes.

The practice of continuing medical education (CME) is expected to enable practicing physicians to apply contemporary methods in the provision of health care. Modern treatment and technological methods are crucial to providing health care. New technologies and protocols are more effective in curing patients and therefore ultimately more cost-effective.

In 2000, the Union of European of Medical Specialists (UEMS) established a body called European Accreditation Council for Continuing Medical Education (EACCME).Its purpose is:

1-Harmonization and improvement of the quality of continuing education in Europe.

2-Provision of nonbiased education to European colleagues according to mutually agreed quality requirements.

3-Guarding of the authority of national Continuing Medical Education (CME) regulatory bodies in the European countries.

4-Linking the national CME regulatory bodies in a system of mutual recognition of accreditation of CME activities.

5-Providing a system in which CME credits obtained abroad in EACCME accredited activities are recognized by the national CME regulatory bodies.

In many countries of the world doctors are under the obligation to participate in continuing professional education according to national guidelines. The need for international standards has been emphasized in Europe.

Clearly, there are differences between the CME programs in various countries. In some the system is mandatory and in others is voluntary. However, in most countries where the CME system is voluntary, physicians participating successfully receive a diploma which can be used as an additional qualification of high standard continuous education in comparison to those who do not participate to the system.

Many are looking forward to reach a state when the measuring system of an individual's participation in CME/CPD activities would be the same for all countries in the world.

An abundance of educational theory, design, and delivery of continuing medical education (CME) learning interventions, including their impact on learners, are described in the health and social sciences literature. However, establishing a direct correlation between the acquisition of new skills by learners and patient outcomes as a result of a planned CME learning intervention has been difficult to demonstrate.

However, in Australia positive relationship between acquisition of a new skill by learners and improved patient outcomes as a result of this planned CME learning intervention has been reported. Table-1 summarizes the general principles and universally or widely accepted requirements and standards for CME/CPD activities.

Table 1: General principles, concepts and universally or widely accepted requirements and standards for CME/CPD activities
I-GENERAL PRINCIPLES AND CONCEPTS
A-The Continuing medical education (CME) Programs has become an integral component of state-of-the-art medical practice.
B-CME is a strategic way to improve the quality of the health system
C-Rehabilitation of outdated knowledge and skills. Retraining on newer and more appropriate methods in diagnosis and treatments allows physicians to provide good quality health care delivery.
D-Ethical obligation The practice of CME has become an ethical obligation to physicians and the disciplinary authority of the profession.
II-REQUIREMENTS
A-Medical education background The integrity of undergraduate medical education should be maintained by adhering to the generally accepted standards of medical school and by national final examination rather than allowing each medical school to conduct their own final examination.
B-Postgraduate residency training and education. The integrity of Postgraduate residency training and education should be maintained by appropriately designed programs and supervised by a qualified residency training committee.
C-CME accreditation body The presence of a qualified independent body of accreditation of CME activities and programs at the national level to guarantee the quality of CME programs and its

independence and to monitor physician participation.
D-Appropriate rewarding systems
The presence of appropriately designed scoring system for CME achievement and appropriate rewarding systems including a system of credits that expresses the professional value of continuing medical education activities.
E-Recertification when appropriate

Acknowledgement: The author is very grateful for Dr Bernard Ferguson, president of the International Association of Medical colleges for his help in the revision of this paper.

References

1-Christodoulou N. Continuing medical education and continuing professional development in the Mediterranean countries. EURA MEDICOPHYS 2007; 43:195-202.

2-Quintaliani G, Zoccali C. [Continuous medical education]. G Ital Nefrol. 2004 Jul-Aug; 21(4):355-61. [Article in Italian]

3-Durning SJ, Hemmer P, Pangaro LN. The structure of program evaluation: an approach for evaluating a course, clerkship, or components of a residency or fellowship training program. Teach Learn Med. 2007 Summer; 19(3):308-18.

4-Routil W. [The system of continuing professional education of medical doctors in Austria. Structure, guidelines and quality management] Bundesgesundheitsblatt Gesundheitsforschung Gesundheitsschutz. 2006 May; 49(5):433-8 [Article in German].

6-Prien T, van Aken H. [Concept for continuing medical education of the European Union of Medical Specialists-- EUMS]. Z Arztl Fortbild Qualitatssich. 1999; 93(8):563-7 [Article in German].

8-.DeLisa JA. Maintenance of certification and pay for performance: implications for physiatry. Am J Phys Med Rehabil 2006; 85:187-91.

9-Bellamy N, Goldstein LD, Tekanoff RA. Support, Non-U.S. Goverment. Continuing medical education-driven skills acquisition and impact on improved patient outcomes

in family practice setting. J Contin Educ Health Prof. 2000
Winter; 20(1):52-61.

Comparison of 1999 and 2003 current cigarette smoking behavior among Jordanian adolescents: the Global Youth Tobacco Surveys.

The New Iraqi Journal of Medicine 2007; 3 (3): 11-18.

Emmanuel Rudatsikira

Departments of Global Health, Epidemiology and Biostatistics   School of Public Health, Loma Linda University, Loma Linda, California, United States of America.

Imad Al-Doghim

Jordanian Royal Medical Services, Amman, Jordan.
Adamson S. Muula

Department of Community Health, University of Malawi, College of Medicine, Blantyre, Malawi

Seter Siziya

Department of Community Medicine, University of Zambia, School of Medicine, Lusaka, Zambia

Abstract

Background: Adolescent cigarette smoking has received particular attention in the past two decades. Comparison of prevalence estimates trends is likely to inform public health intervention strategies. This study was conducted to compare the prevalence of current cigarette smoking among school going adolescents in Jordan between 1999 and 2003.

Methods: Cross sectional, questionnaire-based study among school going adolescents in the Jordanian Global Youth Tobacco Survey 1999 and 2003.

Results: The overall prevalence of smoking in 1999 was 16.9% (95% CI 15.7%-18.1%) versus 15.5% (95% CI 14.5-16.5) in 2003. In terms of gender distribution 26.9% (95% CI 24.5%-29.3%) males were current smokers in 1999, while 20.0% (18.4%-21.6%) were smokers in 2003. 12.4% (95% CI 10.3%-14.5%) females were smokers in 1999 and 10.1% (95% CI 8.9%-11.2%) females were smokers in 2003. Thus comparing the 1999 estimates to the 2003 suggests that there has been an overall drop in prevalence of current smoking among school going adolescents in Jordan.

Conclusion: Widespread antismoking public health interventions may have resulted in the observed reduction in current smoking prevalence in Jordan between 1999 and 2003. There is need to continue monitoring the trends in smoking among adolescents.

Keywords: adolescents, cigarette, smoking, Jordan, non-communicable diseases.

Introduction

Tobacco is the single most important preventable cause of cardiovascular morbidity and cancers in the world [1-4]. Many adult smokers initiate the habit as adolescents. Adolescent smoking is also of public health significance as smoking is a marker of many other harmful lifestyles [5-7].

Akpinar et al [8] have reported prevalence of 26.6% among 13 year olds and 43.7% among 17 years olds in Turkey. In Syria, prevalence of adult cigarette smoking was reported at 56.9% and 35.3% among adult males and women [9]. The Centers for Disease Control and Prevention has reported that 21% and 2.1% of male and female adolescents in Kurdistan Region of Iraq were smokers in 2005 [10].

Much of the data on cigarette smoking in Jordan among young people have mostly come from studies in which study participants were university students or patients attending dental services [11, 12]. While these studies could be representative of their respective source populations, they are unlikely to be representative of the adolescent general population.

Two waves of the Global Youth Tobacco Survey (GYTS) were done in Jordan (1999 and 2003) using the GYTS standard methodology [13-14] to estimate the prevalence of tobacco use and associated factors among school-going adolescents. The GYTS Collaborating Group and Warren et al. have reported prevalence of current smoking among adolescents in 1999 in Jordan as 16.6% [15, 16]. While cross national comparisons of the prevalence of adolescent cigarette smoking have been conducted before [14, 17, 18], there is little information about same country comparison over the years. The objective of this study was to compare the prevalence of cigarette smoking among school going adolescents in Jordan in 1999 and 2003. We also assessed whether there had been changes in the number of cigarettes smoked per day by adolescents who were current cigarette smokers.

Methods

Our study was secondary analysis of the data obtained from the Jordanian Global Youth Tobacco Survey. A comprehensive description of the GYTS methodology has been reported elsewhere [20-21]. In brief, the GYTS is a cross sectional school-based survey of students aimed at ages 13–15 years. Two-stage sample design strategy is used in which schools are selected proportional to their enrollment size. Within any selected school, random selection of classes is done. All students within the selected classes are eligible to participation regardless of their actual ages. A standardized questionnaire is self-completed anonymously by the students and it takes students between 30 to 40 minutes to complete. The questionnaire is aimed at collecting the following information among young people: cigarette smoking and other tobacco use; knowledge and attitudes of towards cigarette smoking; role of the media and advertising and their use of cigarettes; access to cigarettes; tobacco-related school curriculum; exposure to environmental tobacco smoke (ETS) and cessation of cigarette smoking. For the purpose of this study however, only data related to estimation of prevalence of current cigarette smoking, gender distribution of current smoking and number of cigarettes smoked per day will be reported.

Current smoking is defined as having ever smoked even one puff in the past 30 days. The question asked was: "During the past 30 days, have you smoked part or all of a cigarette?" The number of cigarettes smoked was assessed by asking the question: "During the past 30 days (one month), on the days you smoked, how many cigarettes did you smoke?"

A weighting factor was used in the analysis to obtain prevalence of the outcome to reflect the likelihood of sampling each student and to reduce bias by compensating for differing patterns of non response. The weight used for estimation is given by the following formula: $W = W1 * W2 * f1 * f2 * f3 * f4$, where $W1$ = the inverse of the probability of selecting the school $W2$ = the inverse of the probability of selecting the classroom within the school $f1$ = a school-level non response adjustment factor calculated by school size category (small, medium, large) $f2$= a class-level non response adjustment factor calculated for each school $f3$ = a student-level non response adjustment factor calculated by class $f4$ = a post stratification adjustment factor calculated by grade.

Ethical considerations

Permission to conduct the study was obtained from the relevant authorities within the Ministries of Health and Education. Participation by the eligible students was voluntary. Data collection was conducted in school by trained assistants and questionnaires were administered without the presence of their teachers.

Data Analysis

Data were analyzed using SUDAAN 9.0 (Research Triangle Institute, Research Triangle Park, Durham, North Carolina, USA). Proportions and 95% confidence intervals (CI) were obtained as estimates of prevalence.

Chi-square tests were used to compare the proportions. An α value was set at 0.05 and so p value of <0.05 was considered statistically significant.

Since the age distribution of study participants between the two survey waves was different, the two study groups were not comparable in this regard. As prevalence of smoking in part is dependent on age, there was need to standardize the estimates before meaningful comparisons could be done. The following formula was used:

$(\Sigma Pj * Aj) / N * 100$

$Pj =$ 1999 age specific prevalence of current smoking for age j

$Aj =$ number of participants in age category j in 2003

$N =$ total number of study participants in 2003

The 100 in the denominator is required when prevalence Pj is reported as %, otherwise of Pj is reported as ranging from 0 to 1, the 100 in the denominator is not required.

Results

3912 adolescents participated in the 1999 survey. Information of gender was available for 3681 (94.1%). Of those who had data available1682 (45.7%) were males and (54.3%) were females in 1999. In 2003, 6313 students participated and information on gender was available for 5838 (92.5%) participants. Of these 2874 (51.2%) were males and 2964 (50.8%) were females. The median age in both 1999 and 2003 surveys was 14 years. However, the age distribution of study participants between the two survey years was different as shown in table 1.

Age	Frequency (%)	
	1999	2003
11 or younger	353 (8.9)	403 (7.2)
12	472 (11.8)	295 (5.5)
13	1043 (25.0)	520 (9.2)
14	1059 (26.5)	1640 (26.5)
15	745 (22.1)	1523 (25.2)
16	141 (4.3)	1414 (23.4)
17 or older	49 (1.4)	189 (3.1)
Total	3862	5984

Table 1: Age distribution of study participants in 1999 and 2003

Prevalence of current cigarette smoking .The overall prevalence of smoking in 1999 was 16.9% (95% CI 15.7-18.1) versus 15.5% (95% CI 14.5-16.5) in 2003. In terms of gender distribution 26.9% (95% CI 24.5-29.3) males were current smokers in 1999, while 20.0% (18.4-21.6) were smokers in 2003. 12.4% (95% CI 10.3-14.5) females were smokers in 1999 and 10.1% (8.9-11.2) females were smokers in 2003.

Age-specific current cigarette smoking

We also aimed to compare the age-distributed current cigarette smoking prevalence between 1999 and 2003. The results are reported in table 2.

When the 1999 overall prevalence (18.7%) was standardized to the 2003 age distribution, the prevalence was 22.0% and was statistically significantly different from the 2003 estimate (p<0.01).

We also aimed to assess whether there has been a change in the number of cigarette smoked on each smoking days by the adolescent smoker. There was a statistical difference in the number of cigarettes smoked per day between the 1999 and 2003 surveys as reported in table 3. However there was no general pattern observed.

Age category in yrs	1999 Prevalence	2003 Prevalence	p value
11 yrs or younger	19.3	20.8	0.08
12	19.6	34.1	<0.01 *
13	13.8	20.5	0.08
14	15.8	14.2	0.02*
15	20.6	20.5	0.45
16	36.2	24.9	<0.01 *
>=17	42.6	35.9	0.54

Table 2: Age-specific current cigarette smoking prevalence among adolescents in Jordan 1999 and 2003 (*statistically significant at α= 0.05).

Number of Cigarettes per day	1999 Prevalence in %	2003 Prevalence in %	p-value
< 1	46.6	28.3	<0.01
1	28.0	24.8	<0.10
2 to 5	13.9	22.4	<0.01
6 to 10	6.2	11.6	<0.01
11 to 20	3.7	6.4	<0.01
>20	1.7	6.6	<0.01

Table 3: Number of cigarette smoked per day among current adolescent cigarette smokers in Jordan 1999 and 2003

Discussion

The prevalence of current cigarette smoking among school-going adolescents in Jordan was 16.9% in 1999 and 24.6% in 2003. Our 1999 current smoking prevalence estimate is different from the 16.6% reported by the Global Youth Tobacco Survey Collaborative Group (GYTSG) in 2002 using the same data [15]. The reason for the difference was that the GYTSG calculated the prevalence figure as number of participants reporting current smoking divided by total number of study participants. In our estimate, the numerator was the same as in the GYTSG but the denominator was only those who completed the question on current smoking. In the 1999 survey 172 (4.4%) did not the answer the question: "During the past 30 days, have you smoked part or all of a cigarette?" Including these study participants within the denominator was similar to assuming that they were all non-smokers. We chose to use only complete case analysis, and so our estimate is slightly higher, although the difference with the GYTSG was not statistically significant.

The prevalence of smoking in 1999 at 16.9% was only slightly higher than 15.5% estimated in 2003, p= 0.40. However if the 1999 prevalence was age-standardized to the 2003 sample, the prevalence of cigarette smoking rises to 22.0% (p<0.01). This suggests that the prevalence estimate in the earlier sample was statistically different from the 2003 sample, if the age distributional differences between the samples are considered.

The Centres for Disease Control and Prevention (CDC) has reported an overall current smoking prevalence among school going adolescents in the Eastern Mediterranean region as 15.3% [14]. Our estimates are not far different from the 'average' prevalence for the region. This suggests that most countries in the region have relatively high prevalence of smoking.

The decrease in current smoking prevalence in Jordan could be explained, partly if not to a great extent by the public health interventions that the Kingdom of Jordan has taken over the past several years [22].

In 1998, the Jordanian government put in place anti-smoking legislation to prohibit smoking in public places, it has banned the advertisement of cigarettes in the media, and a national committee has been established to draw-up strategies and programs to combat smoking. In 2001, the government used postage stamps to carry adolescent relevant antismoking messages. The committee on the prevention of smoking in Jordan has wide representation including Health Ministry, UNICEF, the Public Security Department, Jordan Bar Association, the National Anti-Smoking Society and the Ministry of Awqaf and Islamic Affairs.

We also aimed to assess whether there was a change among current cigarette smokers with regard to number of cigarettes smoked per day. The proportion of smokers smoking less than a cigarette a day to 5 cigarettes a day were higher in 1999 as compared to 2003. However the proportion having at least 6 cigarettes per day were significantly higher in 2003. This may suggest that although overall smoking prevalence in 2003 was lower than in 1999, the proportion of smokers taking higher numbers of cigarettes increased over the period. Considering that there is an increasing dose-response association between the number of cigarettes and adverse health outcomes, a significant proportion of adolescents were exposing themselves to greater potential harm in 2003. It is also important to recognize that adolescents who smoke cigarettes may also be engaging in other unhealthy behaviors [23]. A holistic approach to overall healthy living is therefore likely to be of greater public health significant than individual behavior approach.

This study has a number of limitations. The data used were obtained through self-completed questionnaire. There is a possibility that some study participants may have misreported their exposure status. However, Brener et al has assessed a method of data collection the GYTS in the United States and they have reported high reliability [24].

While reliability of data collection instrument was acceptable in the United States, we are not aware as to whether the methodology also has high reliability in the Eastern Mediterranean region. We suggest that future studies assess such issues.

Also by asking how many cigarettes the adolescent smoked on the days s/he smoked assumed that the adolescent would smoke a regular number of cigarettes each day. This may result in mis-classification based due to recall problems. It is also not known whether all or some study participants responded thinking that the researchers were asking for the 'average' number of cigarettes per smoking day. Obviously some study participants may have smoked more on certain days and not much on others.

Another limitation of the GYTS methodology is that self reported history of current smoking is not verified by biomarkers such as salivary or blood cotinine level or exhaled carbon monoxide [25-27]. We suggest that the Global Youth Tobacco Survey Collaborating Group consider validating the questionnaire with biomarkers in future surveys. As the GYTS methodology only allows recruitment of students who are present in school on the day that the survey is administered, our findings may not be applicable to all students. Also the findings may not be applicable to out-of school adolescents.

In between the two initiatives, the Jordanian National Anti-Smoking Strategy has been launched [28]. It is expected that such initiatives will support the reduction in the prevalence of smoking among all age groups in Jordan.

Conclusion We find that the overall prevalence of current cigarette among adolescents who participated in the Jordanian Global Youth Tobacco Survey in 1999 was much higher than that obtained in 2003. We suggest that public health interventions aimed to prevent smoking in Jordan may have started bearing fruit.

Conflict of Interest The authors declare no conflict of interest.

Acknowledgements The GYTS is a collaborative project of WHO/CDC/participating countries. Analyses of GYTS data are not necessarily endorsed by the WHO/CDC/participating countries. We are grateful also to Drs Mohammed Shreim and Ayub Hiba for coordinating the Global Youth Tobacco Surveys in Jordan.

References

1-Ansary-Moqhaddam A, Huxley R, Barzi F et al. The effect of modifiable risk factors on pancreatic mortality in populations of the Asia Pacific region. Cancer Epidemiol Biomarker Prev 2006; 15: 2435-40.

2-Brand RM, Jones DD, Lynch HT, Brand RE, Watson P, Ashwathnayaran R, Roy HK. Risk of colon cancer in hereditary non-polyposis colorectal cancer patients as predicted by fuzzy modeling: influence of smoking. World J Gastroenterol 2006; 12: 4485-91.

3-Kaur J, Bains K. A study of the risk factor profile of cardiovascular diseases in rural Punjabi male patients. Indian J Public Health 2006; 50(2):97-100.

4-Toustad S, Andrew-Johnston J. Cardiovascular risks associated with smoking: a review for clinicians. Eur J Cardiovasc Prev Rehabil 2006; 13(4): 507-14.

5-Rudatsikira E, Siziya S, Kazembe LN, Muula AS: Prevalence and associated factors of physical fighting among school-going adolescents in Namibia. Ann Gen Psychiatry 2007; 6:18.

6-Dierker LC, Sledjeski EM, Botello-Harbaum M, Ramirez RR, Chavez LM, Canino G: Association between psychiatric disorders and smoking stages within a representative clinic sample of Puerto Rican adolescents. Compr Psychiatry 2007; 48:237-44.

7-Haddad LG, Malak MZ. Smoking habits and attitudes towards smoking among university students in Jordan. Int J Nurs Stud 2002; 39: 793-802.

8-Akpinar E, Yoldascan E, Saatci E. The smoking prevalence and the determinants of smoking behaviour among students in Cukurova University, Southern Turkey. West Indian Med J. 2006; 55:414-9.

9-Ward KD, Eissenberg T, Rastam S, Asfar T, Mzayek F, Fouad MF, Hammal F, Mock J, Maziak W. The tobacco epidemic in Syria. Tob Control. 2006; 15 Suppl 1:124-9.

10-Centers for Disease Control and Prevention (CDC). Tobacco use among students aged 13-15 years--Kurdistan Region, Iraq, 2005. MMWR Morb Mortal Wkly Rep. 2006 26; 55:556-9.

11-Alomari Q, Barrieshi-Nusair K, Said K. Smoking prevalence and its effect on dental attitudes and behaviour among dental students. Med Princ Pract 2006; 15: 195-9.

12-Kyrlesi A, Soteriades ES, Warren CW, Kremastinou J, Papastergiou P, Jones NR, Hadjichristodoulou C. Tobacco use among students aged 13-15 years in Greece: the GYTS project. BMC Public Health. 2007; 7:3.

13-Arora M, Reddy KS. Global Youth Tobacco Survey (GYTS) - Delhi. Indian Pediatr 2005; 42:850-1.

14- Centers for Disease Control and Prevention (CDC). Use of cigarettes and other tobacco products among students aged 13-15 years--worldwide, 1999-2005. MMWR Morb Mortal Wkly Rep. 2006; 55:553-6.

15- Global Youth Tobacco Survey Collaborative Group. Tobacco use among youth: a cross country comparison. Tob Control 2002; 11:252-70.

16-Warren CW, Riley L, Asma S, Eriksen MP, Green L, Blanton C, Loo C, Batchelor S, Yach D. Tobacco use by youth: a surveillance report from the Global Youth Tobacco Survey project. Bull World Health Organ 2000; 78:868-76.

17-Global Youth Tobacco Survey Collaborating Group. Differences in worldwide tobacco use by gender: findings from the Global Youth Tobacco Survey. J Sch Health 2003; 73:207-15.

18-Muula AS, Mpabulungi L. Cigarette smoking prevalence among school-going adolescents in two African capital cities: Kampala Uganda and Lilongwe Malawi. Afr Health Sci 2007; 7:45-9.

19-Global Tobacco Surveillance System Collaborating Group. Global Tobacco Surveillance System (GTSS): purpose, production, and potential. J Sch Health. 2005; 75:15-24.

20-Centers for Disease Control and Prevention (CDC). Use of cigarettes and other tobacco products among students aged 13-15 years--worldwide, 1999-2005. MMWR Morb Mortal Wkly Rep 2006; 55: 553-6.

21-Rudatsikira E, Abdo A, Muula AS. Prevalence and determinants of adolescent tobacco smoking in Addis Ababa, Ethiopia. BMC Public Health 2007; 7:176.

22-Kandela P. Jordan starts campaign to tackle high rates of smoking. Lancet 2000; 355: 1800

23-Miller JW, Naimi TS, Brewer RD, Jones SE. Binge drinking and associated health risk behaviours among high school students. Pediatrics 2007; 119: 76-85.

24-Brener ND, Kann L, McMannus T, Kinchen SA, Sundberg EC, Ross JG. Reliability of the 1999 youth risk behaviors survey questionnaire. J Adolesc Health 2002; 31: 336-42.

25- Hung J, Lin CH, Wang JD, Chann CC: Exhaled carbon monoxide level as an indicator of cigarette consumption in a workplace cessation program in Taiwan. J Formos Med Assoc 2006; 105: 210-3.

26-Jenkins RA, Counts RW: Personal exposure to environmental tobacco smoke: salivary cotinine, airborne nicotine, and nonsmoker misclassification. J Expo Anal Environ Epidemiol 1999; 9: 352-63.

27-Low EC, Ong MC, Tan M: Breath carbon monoxide as an indication of smoking habit in the military setting. Singapore Med J 2001; 45:578- 82.

28-Jordan Times. National Anti-smoking strategy to be announced today. May 2, 2002 accessed on 26 September 2007from
http://www.jordanembassyus.org/05302002006.htm.

Variation in the Lobes and Fissures of the Right Lung: An Anatomical perspective

The New Iraqi Journal of Medicine 2007; 3 (3):19-22.

Azian Abd Latiff
Associate Professor and Head of the Department of Anatomy
Faizah bt Othman
Associate Professor
Farihah Haji Suhaimi
Associate Professor
Srijit Das
Lecturer
Hairi Ghazalli
Senior Medical Laboratory Technologist
Department of Anatomy Faculty of medicine
Universiti Kebangsaan Malaysia
Jalan Raja Muda Abdul Aziz
50300 Kuala Lumpur, Malaysia

Abstract

Background: The right lung is known to exhibit variations regarding the presence of fissures and lobes. Often the anomalies pertaining to the lungs remain undetected if they are asymptomatic. The variations may be detected incidentally in routine autopsies and cadaveric dissections.

Methods: We studied thirty five cadaveric right lungs (n=35) to detect any anomalies pertaining to fissures and lobes.

101

Results: Out of thirty five specimens studied, anomalous fissure and lobe was detected in a single lung specimen. The anomalous lung exhibited a single oblique fissure which did not extend to the inferior border; rather it extended to the anterior border of the right lung. The presence of the single oblique fissure resulted in formation of two lobes in the right lung.

Conclusion: Anatomical knowledge of the abnormal fissures and lobes of the lungs may be important for surgeons performing lobectomies. The presence of the anomalous lobes and fissures may also result in erroneous interpretation of skiagrams.

Keywords: Lung; Lobes; Anomaly; Variation; Anatomy.

Introduction

According to standard anatomy textbook, the right lung has two fissures i.e. the oblique fissure and the horizontal fissure which divides it into three lobes: superior, middle and the inferior [1]. The oblique fissure (OF) can be traced, traversing a downward course thereby meeting the inferior border of the lung at a distance of 7.5 cm behind the anterior end [1]. The horizontal fissure (HF) passes from the OF at the level of the mid-axillary line to the anterior border of the lung at the level of the sternal end of the fourth costal cartilage. An OF, passing to the anterior border of the right lung, is a rare entity.

In the present study, we describe the gross anatomical features of an anomalous lung with an OF traversing a horizontal course to terminate at the anterior border of the right lung, thereby resulting in the formation of unusual two lobes instead of the usual two lobes. Anatomical knowledge of the normal and abnormal fissures and lobes of the lung may be important for surgeons and radiologists in day to day clinical practice and the present study is made to highlight such.

Materials and Methods

We observed thirty five right lungs (n=35) for the presence of anomalous fissures and lobes. No information regarding the history of the individuals was available. The right lungs were carefully studied, morphometric measurements were taken and the specimen was photographed (Figure.1). The anomalous specimen was also compared to a normal specimen (Figure.2).

Results

Out of thirty five specimens studied, we observed anomalous fissures and lobes in a single specimen (2.84%). The posterior border of the right lung was traced from above and a distance of 13 cm from the apex, an OF was traced to traverse a course towards the anterior border ('OF' in Fig.1).

The OF on its origin from the posterior border was prominent over a distance of 7 cm, thereby becoming faint while reaching the anterior border. The extent of the posterior border from the origin of the OF till the inferior border, measured 12.5 cm. There was only a single fissure i.e. the OF with the HF being absent. Only two lobes i.e. superior and the inferior were observed in the present case.

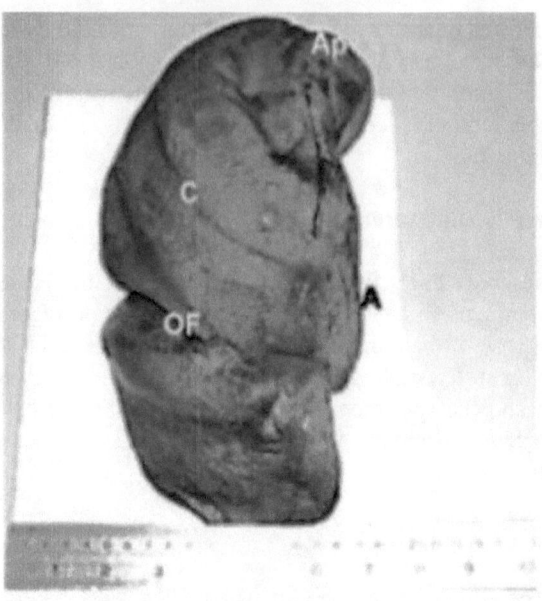

Figure.1: Photograph of anomalous right lung showing:
Ap: Apex; **A:** Anterior border; **C:** Costal surface; **OF:** Oblique fissure

Discussion

The anomalous fissures and the lobes may be due to defective development of the lungs. It has been reported that the fissures are the spaces which separate individual bronchopulmonary buds or segments and there is obliteration except along the two planes which give rise to the HF and the OF [2]. Whenever there is non obliteration of these spaces, the accessory fissure arises [2].

Many past research studies have focused on the accessory fissures. A past research study had reported an accessory fissure in the right lung between the superior and the basal segments of the lower lobes [3]. Studies performed by high resolution CT examinations on 30 healthy individuals found that in 87% and 77% cases, there were incomplete fissures in the right and the left lung, respectively [4]. Incompleteness of fissures (fusion between the lobes) has been reported to have 70% incidence [5].

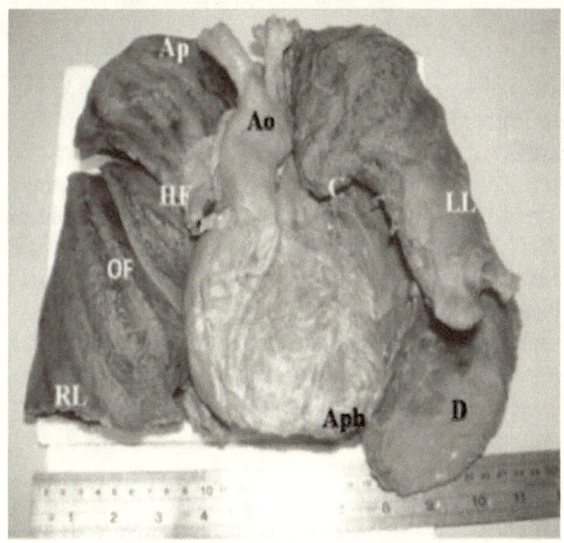

Figure.2: Photograph of normal right and left lungs showing: AP: Apex of lung; **HF:** Horizontal fissure; **OF:** Oblique fissure; **RL:** Right lung; **Ao:** Arch of aorta; **Aph:** Apex of heart; **C:** Cardiac notch; **LL:** Left lung; **D:** Diaphragmatic surface

A past study had even attempted to classify the pulmonary fissures [6]. The grades were classified as :- Grade 1-complete fissures, separate lobes; Grade 2- complete visceral cleft but parenchymal fissure at base of fissures; Grade 3-visceral cleft evident for part of the fissure; Grade 4-complete fusion of the lobes with no evidence of fissural lobes. The same study had defined the pulmonary artery to be centrally located to the OF and termed the displacement of the artery in anterior and the posterior directions as 'imbalance'.

In CT scans, the major fissures may be visualized as lucent band, less often a line and least often as a dense band [7]. Localization of tumor or any mass in the lung along the fissures may be important. Often, the abnormal fissures may result in erroneous interpretation of skiagrams. It has been found that incomplete fissures often give an atypical appearance of the pleural effusion [8]. An incomplete major fissure has been identified to be linked to the spread of any disease and causing collateral drift [8]. Some radiologists have termed it as 'incomplete fissure sign' [8]. In the present case, the abnormal course of the OF will definitely cause an imaging conflict.

An incomplete fissure has been found to be the cause for post operative leakage [6]. It has been found also that accessory fissure acts as a barrier to the infection spread causing a sharply marginated pneumonia which can be mistaken as atelectasis or consolidation [9].

An earlier anatomical study had defined the presence of an accessory fissure in the right lung with the presence of four lobes [10]. Thus, it is an accepted fact that the right lung may exhibit variations regarding the number of fissures and lobes.

Conclusion

Although, presence of two lobes and an accessory 'azygos lobe' in the right lung is not an uncommon finding as described in anatomy textbook [11], the awareness of the same may be beneficial for surgeons performing lobectomies and radiologists interpreting skiagrams. The present anatomical finding is a sincere attempt to highlight such.

References

1-Standring Susan. Gray's Anatomy. The Anatomical Basis of Clinical Practice. 39th ^{edition} 2005, Philadelphia, Elsevier Churchill Livingstone; pp-1067-1070.

2-Meenakshi S, Manjunath KY, Balasubramanyam V. Morphological variations of the lung fissures and lobes. Indian J Chest Dis Allied Sci. 2004; 46:179-82.

3-Aldur MM, Deck CC, Celik HH, Tasçioglu AB. An accessory fissure in the lower lobe of the right lung. Morphologie. 1997; 81: 5-7.

4-Frija J, Naajib J, David M, Hacein-Bey L, Yana C, Laval-Jeantet M. [Incomplete and accessory pulmonary fissures studied by high resolution x-ray computed tomography]. J Radiol; 69: 163-170.

5-Raasch BN, Carsky EW, Lane EJ, O'Callaghan JP, Heitzman ER. Radiographic anatomy of the interlobar fissures: a study of 100 specimens. AJR Am J Roentgenol 1982; 138: 1043-9.

6-Craig SR, Walker WS. A proposed anatomical classification of the pulmonary fissures. J R Coll Surg Edinb 1997; 42: 233-4.

7-Proto AV, Ball JB Jr. Computed tomography of the major and minor fissures. AJR Am J Roentgenol 1983; 140: 439-48.

8-Hayashi K, Aziz A, Ashizawa K, Hayashi H, Nagaoki K, Otsuji H. Radiographic and CT appearances of the major fissures. Radiographics 2001; 21: 861-74.

9-Godwin JD & Tarver RD. Accessory Fissures of the Lung. AJR Am J Roentgenol 1985; 144: 39-47.

10-Modgil Vishal, Das Srijit & Suri Rajesh. Anomalous Lobar Pattern of Right Lung: A Case Report. International Journal of Morphology 2006; 24: 5-6.

11-Moore KL, Agur AMR. Essential Clinical Anatomy. Second edition 2002, Baltimore, Lippincott Williams & Wilkins, pp-77.

Letter to the editor: Sexual dysfunction in uncomplicated diabetic men: Positive association with insulin therapy

The New Iraqi Journal of Medicine 2007; 3 (3): 77-78.

Ghazi Aboud
Specialist psychiatrist
Al Rashad Hospital
Baghdad
Mohammad Rasheed.
Specialist psychiatrist
Iraqi Ministry of Health

Erectile Dysfunction(ED), the inability to achieve and / or to maintain an erection for a sufficiently long period of time to permit satisfactory sexual intercourse has been attributed to diabetic nephropathy in diabetic patients [1, 2].ED showed a positive correlation with age after 65 years, history of diabetes of more than 10 years and was not correlated to the type of diabetes mellitus and diabetic therapy [3, 4, 5].In one study the prevalence of ED is lower in type 1 diabetic patient. [6].Orgasmic dysfunction was correlated positively with the duration of diabetes but not with the type of received by the patients [7]. The prevalence of erectile dysfunction amongst diabetic men in a hospital clinic population has been reported to be between 35 and40 % [8, 9].Also ED has been reported in up to 75% of diabetic men [8-12].

The aim of this paper is to report a positive correlation between ED in small series of relatively health patients with uncomplicated diabetes. From January to march 2003, 28 male diabetic patients attending the outpatient of general hospital in Baghdad, were asked to complete a questionnaire of 11 questions about their disease and sexual problems. All the patients answered the questions completely. We excluded any patient with sexual dysfunction prior to the diabetic illness and other associated medical condition. The age of patients ranged from 19 to 50years (mean 35.5y.)Twenty of them (71%) were married and 8 (29%) were single.

Twelve patients (43%) were diagnosed as diabetic type 1and16 patients (57%) were diabetic type 2 .The diagnosis depended on the international classification of the diabetes mellitus& was done by a specialist in endocrinology.16 patients (57%) were on insulin treatment, 9 patients (32%) on hypoglycemic agents and 3 patients (11%) on mixed treatment.

The answers were analyzed and statistical methods of percentages and chi-square had been used.10 patients (35.5%) complained from ED and one patient (3.5%) complained from lack of interest, while17 patients (61%) did not have sexual dysfunction. Among patients with sexual dysfunction, 5 patients (46%) on insulin therapy, 3 patients (27%) on hypoglycemic agents and the remainder 3 patients (27%) were on mixed treatments.

In this series significant numbers of diabetic men were excluded because of prior sexual dysfunction and other medical disorders.

The lack of sexual interest was described by only one patient (3.5%). It appeared that premature ejaculation was considered to remain intact. The most interesting finding of this study was that half of patients with ED were on insulin therapy which is inconsistent with other studies which found that it is lower in type 1 diabetic patients [2].

References

1-Foresta C, Caretta N, Aversa A, et al. Erectile dysfunction. J Endocrinol Invest 2004; 27(1):80-95.

2-Saenz de Tejada 1 , Goldstein I, Azadzoi K, et al. Impaired neurogenic and endothelium mediated relaxation of penile smooth muscle from diabetic men with impotence. N Engl J Med 1989; 320:1025-30.

3-Fedele D, Coscelli C, Cucinotta D, et al. Incidence of erectile dysfunction in Italian men with diabetes. J Urol 2001; 166(4):1368-71.

4-El-Sakka AlTayeb KA et al. Erectile dysfunction risk factors in non-insulin dependent diabetic Suadi patients. J Urol 2003; 169(3):1043-7.

5-Foresta C, Caretta N, Aversa A, Bettocchi C, Corona G, Mariani S, Rossato M. Erectile dysfunction. J Endocrionol Invest 2004; 27(1):80-95.

6-Klein R, Klein BE, Lee KE et al. Prevalence of self-reported erectile dysfunction in people with long-term IDDM. Diabetes Care 1996; 19:135-41.

7-Wilkinson DG. Psychiatric aspects of diabetes mellitus. Review Article. British Journal. Psychiat 1981; 138:1-9.

8-Dunsmuir WD, Holmes SA. The etiology and management of erectile ejaculatory, and fertility problems in men with diabetes mellitus. Diabetic Med 1996; 13:700-8.

9-Price DO, Malley BP, James MA et al. Why are important diabetic men not being treated? Pract Diabetes 1991; 8:10-1.

10-Broderick GA, Schwartz S. Erectile dysfunction in diabetes. Hosp Pract off Ed 1991; 26:139-42,147-55.

11-Delawter DE. Diabetes and impotence. Md Med J 1990; 39:683.

12-Lustman PJ, Clouse RE. Relationship of psychiatric illness to impotence in men with diabetes. Diabetes Care 1990; 13:893-5.

www.ingramcontent.com/pod-product-compliance
Lightning Source LLC
Chambersburg PA
CBHW022024170526
45157CB00003B/1342